SCIENCE
MAGIC

Also by Ormond McGill

The Secret World of Witchcraft
The Encyclopedia of Genuine Stage Hypnotism
The Art of Stage Hypnotism
Professional Stage Hypnotism
Psychic Magic
21 Gems of Magic
Atomic Magic
How to Produce Miracles
Entertaining with Magic
The Mysticism and Magic of India
Hypnotism and Meditation

SCIENCE MAGIC

101 Experiments You Can Do

Ormond McGill

PRENTICE HALL PRESS

New York   London   Toronto   Sydney   Tokyo

Published in 1986 by Prentice Hall Press
A Division of Simon & Schuster, Inc.
Gulf + Western Building
One Gulf + Western Plaza
New York, NY 10023

Originally published by Arco Publishing, Inc.

PRENTICE HALL PRESS is a trademark of
Simon & Schuster, Inc.

**Library of Congress
Cataloging-in-Publication Data**

McGill, Ormond.
　Science magic.

　Summary: Magic tricks based on the principles of science are grouped into the areas of magnetism, numbers, chemistry, dry ice, and physics.
　1. Science—Experiments—Juvenile literature.
2. Scientific recreations—Juvenile literature.
[1. Scientific recreations. 2. Magic tricks] I. Title.
Q164.M37　1984　　507'.8　　84-414
ISBN 0-668-05849-8 (Cloth Edition)
ISBN 0-668-05853-6 (Paper Edition)

Manufactured in the United States of America

10　9　8　7　6　5

CONTENTS

	Preface	vii
	Author's Note	ix
1	Magic with Magnetism	1
2	Magic with Numbers	30
3	Magic with Chemistry	51
4	Magic with Dry Ice	78
5	Magic with Physics	92
6	Teaching Children the Art of Magic	151
	Index	156

Handle tools, chemicals, and all apparatus carefully, and store safely, out of the reach of children. Read the whole section on every trick carefully before you purchase or attempt to construct apparatus for it.

PREFACE

Science Magic: 101 Experiments You Can Do affirms that magic is learning as it presents a host of mystifying magic tricks based upon the principles of science and related subjects. Learning becomes fun when interest is stimulated by magic. Everyone is instinctively attracted to magic: it provides wonder-filled entertainment and is a delightful hobby. All of the tricks in this book are easy to do, and you don't have to be a professional to perform them.

Magicians were really the first scientists, and magic has been the foundation upon which various sciences have been built. Magic uses principles of acoustics, optics, mechanics, magnetism, mathematics, chemistry, and psychology to produce its effects (usually illusions). By using this book, you will learn more about each of these subjects in a very pleasant and useful way. The scientific principles will stay in your mind because of the applications you have learned and the demonstrations you have seen. Scientific facts combined with the art of magic will broaden your horizons and improve your thinking powers.

ORMOND MCGILL

AUTHOR'S NOTE

Any magical apparatus desired for the tricks in Chapter 1, "Magic with Magnetism," such as Magnetized Playing Cards or Magnetic Table Lifting, can be either constructed by yourself or purchased from Louis Tannen, Inc., 1540 Broadway, New York, NY 10036 or Abbott Magic Mfg. Co., Colon, MI 49040. These firms provide elaborate catalogues listing hundreds of tricks.

The effects described in Chapter 2, "Magic with Numbers," require only a pad and pencil, being entirely of a mental nature. The chemicals used for the effects described in Chapter 3, "Magic with Chemistry," are easily obtained from chemical supply houses such as Van Waters and Kogen, McKesson Chemical Company, Matheson Division of Searle, etc. Firms that supply chemicals are in every city of any size. Checking the yellow pages of your telephone directory under "Chemicals" will supply you with many sources.

Dry Ice (frozen carbon dioxide), Chapter 4, can be obtained from any company in your area handling refrigeration products.

The effects described in Chapter 5, "Magic with Physics," are easily constructed at home, with the exceptions of The Decapitated Princess Illusion and the "Black Art" Magic Act. The latter two illusions would have to be built in a workshop. They are presented in this book primarily to show the workings of the optical principles on which they depend.

1

MAGIC WITH MAGNETISM

Magnetism is so magical that there are tricks performed with magnetism—and magnetism performed with tricks. Here are some fascinating magnetic effects.

Quick Experiments with Magnetism

The Magnet Brush

```
You will need:
  a bar magnet
  iron filings
```

Every magnet has a north and a south pole. To demonstrate this, take a bar magnet and dip it in some iron filings. They will immediately be attracted to one end of the bar magnet. Suppose this to be the north pole of the magnet; each end of the filings not in contact with the magnet will become a north pole, while the ends in contact with the magnet will be south poles. Because the north poles will have a tendency to repel each other, the filings will stand out on the surface of the magnet like bristles on a brush, as illustrated.

The Magnetized Poker

You will need:
  a poker
  an iron hammer

Take a poker and hit it several times with a large, iron hammer. Surprisingly, the poker will become magnetized. You have made a magnet.

The Swinging Magnets

You will need:
two pieces of iron wire
thread
a bar magnet

As shown in drawing *a* below, suspend two short pieces of iron wire from threads, so they will hang in contact in a vertical position. Now bring the north pole of a bar magnet close to the wires, and they will separate from each other (as in drawing *b*). This separation of the hanging bits of iron will increase as the magnet approaches them, but there will occur a critical distance at which the attractive force overcomes the repulsive force, causing the wires to converge toward each other (as in

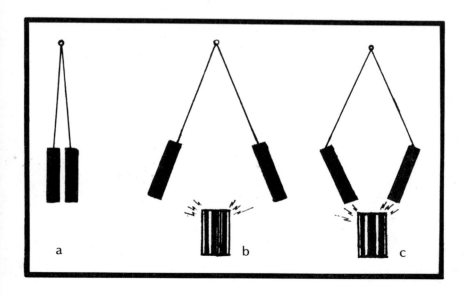

a   b   c

drawing *c*). This is caused by the attractive force of the north pole of the magnet overcoming the repulsive force of the south pole while the north pole still exhibits mutual repulsion.

The Levitated Needle

You will need:
  a bar magnet
  a sewing needle
  thread

Another interesting demonstration of magnetism may be performed by arranging a magnet to stand upright on a table. Then bring a small sewing needle—attached to a length of thread—within the field of the magnet, while keeping hold of the thread to prevent the needle from attaching itself to the magnet. The needle will endeavor to fly to the magnet, but, being prevented from doing so by the thread, will remain curiously suspended in the air. It appears much like a form of magical levitation.

The Magnetic Motor

You will need:
  iron wire
  a cork
  brass wire
  a needle
  a horseshoe magnet
  a wooden block
  a candle or other
    small source of
    heat

You can use this same principle to make a Magnetic Motor. To make this, construct a lightweight wheel about four inches in diameter out of iron wire. For the hub of the wheel, use a cork, and use four brass wires for the spokes. Suspend this wheel on a large, upright needle so that it will turn freely. The arrangement is shown in the drawing on page 6.

Next, arrange a horseshoe magnet a little distance from the wheel on a horizontal support of some kind, and on the same plane as the wheel. The wheel is now in equilibrium before this magnet, for the two branches of the magnet (north and south poles) bear on two equal portions of the wheel's circumference.

Now, using a cigarette lighter, a candle, or a small lamp, heat the portion of the wheel that is near one of the branches of the magnet (either pole). The wheel will begin to turn slowly and continuously as the heated portion tries constantly to get away from the magnet.

The scientific principle involved here is that a magnet attracts iron at an ordinary temperature, but when the iron is heated, it is no longer attracted. Therefore, the cold part of

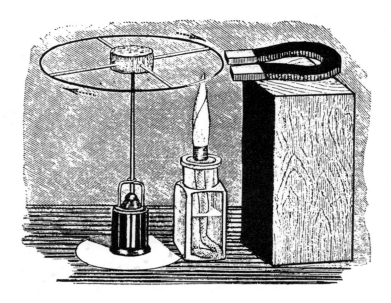

the wheel is attracted by the magnet more than the hot part, and the wheel begins to revolve in the direction indicated by the arrow around and around and around. You have made a Magnetic Motor.

The Dancing Doll

You will need:
  a cork
  7 needles
  a matchbox cover
  a bar magnet
  adhesive tape
  a glass tumbler

The Effect:

This clever trick uses magnetism as its modus operandi (its means of operation). A small doll carved from a cork with a needle through its center as a spindle (as shown in the drawing on page 8) is balanced on the end of a matchbox cover. As you whistle at the doll, it begins to rock back and forth, and "dances" in time to your tune.

Here's How You Do It:

Carve a small doll about one and a half inches high from a cork. Take six steel needles, and, by stroking them in one direction with a magnet, magnetize them. Push these magnetized needles inside of the cork doll, out of sight. Paint a cute face on the doll and decorate it any way you wish. Then insert a large needle crosswise through its center to form a spindle so that when the spindle rests on the end of a matchbox cover, the doll is balanced in an upright position (see the drawing on page 8).

Now, if a small bar magnet held end-on toward the doll, about one foot away, is waved slowly up and down, the doll

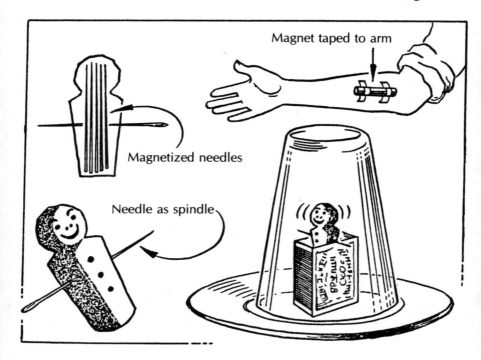

Magnet taped to arm

Magnetized needles

Needle as spindle

will start to swing back and forth in response to the movement of the magnet. Practice this a little, because the speed of the waving of the magnet must be adjusted in time to the doll, so that it adds momentum to the swing of the doll each time. Once the doll is swinging strongly, a magnet now waved out of time with the doll's movements opposes the swinging, and the doll will rapidly slow down and stop moving.

You can get a bar magnet of the right size to be used in this trick at most toy shops. Fasten it to the inside of your arm with adhesive tape, on the bare forearm near the elbow, as shown in the above drawing. Then put your coat or shirt on so that the magnet is concealed.

To perform the trick, after the little doll has been examined by a member of the audience, place it upright in the matchbox cover, and as you whistle wave your hands at it in magic-like passes, and then blow at the doll surreptitiously to start it moving. Let the spectators catch you in the act. It increases the mystery when they think they know how you make the doll dance by blowing at it for now you mystify them completely.

Place the doll in the matchbox on a plate and place an inverted glass over it so that moving the doll by blowing is obviously impossible. Start whistling and waving at the doll under the glass, and it will magically start moving again. At this point, go into a full dancing routine—the glass will make no difference—and, as you whistle, move your arms and hands in time with the tune. Thanks to the concealed magnet attached to your arm, the doll will swing and sway on the matchbox cover, to everyone's amazement. To make it stop "dancing," just wave your arms out of time with its rhythm, and it will slow down and stop.

This little doll will dance via the concealed magnet forever if, when you are not using it, you store it away close beside the bar magnet to maintain the magnetism in the needles.

The Revolving Pencil

You will need:
  a nail
  a magnet
  a new wood pencil

The Effect

This trick is similar to the foregoing cork doll trick in that an object is made to move mysteriously; but, while in the case of "The Dancing Doll" you aim to keep the use of magnetism secret, for this effect you give magnetism credit for doing something magnetically impossible. For the effect, balance the nail carefully—right at its center point of balance—on the edge of your table. Then by carefully weaving the magnet over it, you can cause it to rotate and finally topple off of the edge of the table. Everyone knows that this is possible for iron is attracted by magnetism. But you now offer to perform the same demonstration using a wood pencil, and that is normally impossible.

This time you balance the pencil at its center on the edge of the table, exactly as you balanced the nail previously. And weaving the magnet over it, you cause it to rotate and finally topple off the table.

Here's How You Do It:

As you weave the magnet over the balanced pencil on the edge of the table, you secretly blow upon the pencil. Do this unobtrusively. The stream of breath will cause the pencil to revolve on its balanced center point, and finally cause it to topple off the table. As the attention of the spectators is on the magnet in the trick, the fact that you are blowing on the pencil goes completely unnoticed, and the effect of the wood pencil revolving is most mysterious.

The Magic Arrow

You will need:
  a piece of paper
  a needle
  a cork
  a glass tumbler
  wool cloth
  a nickel
  a match
  a pocket comb

The Effect:

Static electricity and magnetism are closely enough allied in their effects to blend nicely in a magical effect. In this one, you make a "Magic Arrow" by folding a paper in half and in quarters, then opening it and cutting along the creases to form a cross-like arrow, as shown below. Use a paper about three inches by one and one-half inches in size. Cut out the arrow to be approximately two and one-half inches long and one inch wide.

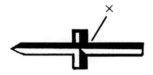

Next, drive the head of a needle into a cork so that the point stands upright from the top of the cork, and place this arrow at its central part (at the meeting of the two folds) gently on the point of the needle, where it will balance. Now, invert a glass over the suspended arrow.

Rub the sides of the glass with a woolen cloth, and the static electricity generated will cause the arrow to turn around until its point stops opposite the part of the glass that is rubbed.

Here's How You Do It:

This is one of the rare tricks where the effect and the modus operandi are one and the same. Just do it as described, and it works. It is an interesting way to demonstrate the scientific fact that the glass becomes electrified by rubbing, which causes it to attract light bodies such as the point of "The Magic Arrow."

The stunt can be made more of a challenge by resting a paper match on the edge of a nickel; you can stand the nickel on its edge by using masking tape to hold it upright, being careful to leave the top edge free. You can then balance the match on the edge of the nickel easily. Now place an inverted glass over the whole arrangement, and challenge anyone to make the match move without touching the glass or moving the table. The spectators will soon decide that the feat is impossible.

To do it, take your pocket comb and run it through your hair several times. Then hold the comb near the glass, but do not touch the glass with the comb. The static electricity generated will attract the match, and it will move.

The "Animal Magnetism" Motor

You will need:
  a piece of paper 3 inches
    × 3 inches
  a long needle
  a cork

Almost everyone has heard of "animal magnetism," which is said to be an energy force generated by some people that has the quality of a magnet. Some people believe in it and others do not. From the standpoint of magical entertainment, this is excellent, because it adds to the mystery, and some of the best tricks performed with "magnetism" are effects that apparently demonstrate this strange ability to exert "animal magnetism."

The Effect:

As in the "Magic Arrow" trick, you fold a piece of paper to make it into the "Magic Motor" in order to demonstrate this strange force. Holding your cupped hands about the "motor," you cause it to move mysteriously, first in one direction and then in another.

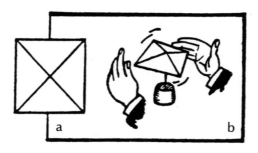

Here's How You Do It:

Take a piece of paper about three inches square, and fold it diagonally from corner to corner. Then open it and make another diagonal fold so that there will be two folds (creases) forming intersecting diagonals. Open the paper, which will now look like a low, partially flattened-out pyramid, as shown in drawing *a* above.

Now, take a long needle and force it through a cork so the point extends upward. Place the cork supporting the point-up needle on your table. Then take the folded paper square and balance it—where the creases intersect—on the point of the needle, placing it so that the four sides of the pyramid point downward. You have here created a very delicate instrument that will spin on the point of the needle at the slightest provocation.

Next, as shown in drawing *b*, place your hands around the needle-supported paper in a semi-cupped position, keeping the hands an inch or so away from the paper so that it can revolve freely. Tell the spectators that you are going to concentrate on the instrument (you can call it your "Animal Magnetism Motor"), and, by your magnetic powers of concentration, make the "motor" operate by causing the "animal magnetism" to emanate from your hands and make the paper twirl.

Hold your hands steady and actually concentrate on the "motor" moving, and it will start to move. At first, it will wobble and perhaps revolve in one direction or the other. As you continue, it will soon be turning rapidly upon the needle point.

Scientifically, heat waves from your hands account for the action, but keep this secret to yourself and let the spectators wonder. In this experiment, you have seemingly demonstrated a mysterious "force" that you will further demonstrate in the following tricks in magical magnetism.

The Human Magnet

> You will need:
>   a pencil or stick
>   or table knife

The Effect:

Everyone knows that a magnet will attract iron to it and hold it, supported by an invisible force. Thus, a main effect of magical magnetism is for the performer to cause his hands to attract and hold objects like a magnet. One of the simplest and undoubtedly one of the oldest of such effects is this one. Take a pencil or a stick or a table knife and grip it firmly in your right hand. Turning your hand downward, support your right wrist with your left fingers, explaining that you must keep the hand absolutely steady for this effect. On opening the fingers of your right hand, the stick or knife magically clings to the undersurface of the hand as though supported by a magnet.

Here's How You Do It:

The drawing on page 17 (*a* and *b*) shows both the effect and the method. In the action of supporting your right wrist, the forefinger of the left hand is secretly extended and supports the stick against the under surface of the right hand when the fingers are opened.

The illusion is good; keep your fingers low so no one can see the palm of your right hand, and perform the action deftly. Just hold the "magnetic" action long enough for the spectators to see it, then remove the concealed forefinger replacing it

along with your other fingers which grip the right wrist, and let the stick, pencil, knife, or whatever is being supported, fall to the floor.

The Magnetized Cane

You will need:
  a chair
  a cane
  black thread

The Effect:

Take a seat in a chair and place the tip of a cane on the floor in front of you. On removing your hands very carefully, the cane is seen by your audience to remain standing upright; and, as you move your hands "magnetically" above it, the cane dances and performs a variety of movements.

Here's How You Do It:

The above drawing shows the secret, which is a length of thin, black thread fastened between the legs just below the knees. Taking a seat in a chair, the cane can be mysteriously supported by the thread and will stand upright. A little movement of the legs, as your hands gesture above the cane, will cause it to dance.

In artificial light, with the spectators seated at a little distance, the black thread is absolutely invisible, and the effect uncanny. The audience is distracted by your hand movements over the cane. To conclude the trick, you have only to stand up from the chair, and pass the cane out for inspection. When you stand, the thread dangles invisibly against your trousers.

A Complete Magical Magnetism Act

You will need:
  newspaper
  bits of colored tissue paper
  a pencil
  a small needle
  a cane
  a small stick or pencil
  a coat with pockets
  an empty cigar box
  a pack of playing cards
  a finger ring
  a small piece of stiff plastic
  a small, lightweight table
  a nail with a head
  a handkerchief or thin scarf

The tricks that follow are assembled into a complete routine that you can have fun with in presenting a Magical Magnetism Act. You will find this a very effective pseudoscientific magical demonstration of "personal magnetism."

The Effect:

What makes an act is not the individual features so much as it is the manner in which these features are grouped together. This is called "routining." Here, the routine consists of tearing strips from a newspaper, magnetizing them and using them to illustrate magnetic attraction and repulsion. The magnetized strips are made to adhere to walls, clothes, table legs, and to various spectators.

Next, show a small stick to the audience. It is placed flat on the palm of your left hand, and by "magnetism" you make it rise and stand upright; then you "make" it descend again slowly to your palm.

The stick is then held by the fingertips at one end, and its other end is rested on the back of a spectator's hand. He or she experiences a succession of definite impulses from the stick. This is repeated with other spectators, as desired. When the end of the stick is placed against an empty cigar box, *raps* are heard as the result of the "magnetism" in the stick.

The stick is next caused to adhere to your inverted palm, and a spectator is allowed to remove it. Then you show a pack of cards. You place your hand flat on the surface of the table and you have a spectator push the cards between the surface of the table and the surface of your palm. When you lift your hand, the cards adhere magnetically to it. Suddenly, the cards scatter and drop to the table. They may be examined.

Climaxing the routine, you place your hand flat on the top of a small table. On lifting your hand into the air, the table clings to the palm as though clinging to a magnet, as you walk with the suspended table among the spectators.

Returning to the front, you release the "magnetism," and the table is placed back in its original position. The act is over.

Here's How You Do It:

There are five tricks used in sequence in this act. Each will be explained in turn.

1. The Magnetized Strips of Newspaper

This stunt actually operates using static electricity. The newspaper used must be dry and the air of the room in which the feat is demonstrated free from humidity. From the newspaper, tear strips about three and a half inches wide and twelve inches long. Take one strip by the end, holding it with the left hand so the strip rests on the front of your left pants leg. With the flat palm of your right hand, give the strip several rapid downward strokes. Actually, only the first two or three strokes are made with the flat palm, then secretly you curl the tips of your right fingernails in contact with the strip as you rub it. The nails will always magnetize the paper, but the flesh will not do so if the least moisture is on the hands. This is a subtle bit of business, and should anyone try to duplicate the stunt by rubbing the paper, they will not be able to do so unless they know about the fingernail technique.

Having magnetized the strips of newspaper, you can place one strip against your pants leg, and it will adhere there. Take another magnetized strip and place it on another part of your body, and it will stay there. A couple of magnetized strips are taken and held together; hold them at the upper end in front of you and away from the body. If properly magnetized, the lower ends on being released will be repelled and will spread widely apart. Introduce your free hand between the two strips, and they fly back to that hand. Remove your hand and the strips again separate. Repeat this several times.

Using a charged strip of newspaper as you would a magnet, several effects can now be performed. Bits of brightly colored tissue paper can be picked up by the strip. A pencil, standing balanced upon one end, can be caused to topple over when the strip is brought near. A cane, balanced at its center on a chair back, can be caused to revolve in either direction as it follows the strip near to one side of either end. In the dark, a tiny electric spark will be produced when someone's finger is touched to the charged strip.

Following these preliminary demonstrations, take fifteen or twenty strips, holding the whole bunch against your pants leg from one end. Magnetize the top strip (as has been explained), peel it off and stick it on a table drape. Magnetize another and put it on the table leg or on the wall. It will hang there magnetically. Then go among the spectators and plaster strips over their bodies—on heads, legs, clothing, etc. This demonstration will cause much laughter as the recipients remove the strips thus acquired and try it themselves, some succeeding and some failing. Some people even jump and scream as they feel a small electric shock when a fully charged strip touches their face, the nose being particularly sensitive.

2. The Magnetized Stick

The second item in the routine, "The Rising Stick," is accomplished in this manner:

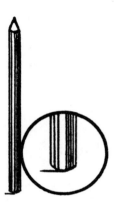

The stick used should be above six inches long and the diameter of a pencil. You can use a quarter-inch piece of dowling for this or a pencil. One end of the stick should be squared, and the other rounded slightly; and a very small needle point should be fixed at right angles to this end. (The drawing above shows how this is done.) Also an unprepared duplicate or pencil (as the case may be) should be ready in your coat pocket.

As the stick is placed on the palm of the left hand, the tiny needle point is inserted in the surface skin of the fleshy part of the rear of the palm. Be careful in doing this so as not to hurt yourself. Remember the needle point goes only into the surface skin. The top end of the stick rests on the slightly curled fingers of the hand. As the fingers of the right hand make "magnetic passes" over the stick, the left hand is very slowly stretched out into more of a flattened position. This action is imperceptible and will cause the stick to rise up from the palm of the hand in a most mysterious manner. Be sure to stretch the hand very slowly, and the deception will be complete. Slowly relaxing the hand causes the stick to resume its original position flat to the palm.

Then take the stick directly from the palm of your hand and place it in your coat pocket while you rub your hands briskly together as though further magnetizing them. This action completed, reach in your pocket and bring out the stick again. Actually, you bring out an unprepared duplicate stick this time, leaving the one with the needle point behind in the pocket; thus you have switched the sticks and are ready for the next effect.

3. The Magnetic Impulse

The unprepared stick in your hand, you are ready for this next demonstration in the routine, in which you rest the stick against the back of a spectator's hand, and he feels the "magnetic impulses."

To accomplish this effect, rub your thumb on your coat sleeve, apparently to generate magnetism but actually to make your thumb very dry. Grasp the blunt end of the stick with your right thumb and your first and second fingers, palm downwards. Grasp stick at a right angle to your hand. The rounded end of the stick is at the bottom. The blunt end rests firmly against the forefinger, about midway between fingertip and first joint. The second finger is placed on the far side of stick, just below the top, this finger being up against the forefinger. Also, the tip of the thumb rests firmly on stick just below the top, and the fleshy ball of the thumb presses *tightly* against both first and second fingers. (See the drawing on page 25.) Now, by pressing the thumb tip against the fingers still harder with an upward pressure, a slight and almost imperceptible "jump" will be made by the thumb tip which still presses firmly against the stick. This causes an invisible vibra-

tion of the stick to be felt as a "magnetic impulse" upon the back of spectator's hand. Or, if the end of the stick is placed against the surface of an empty cigar box (which acts as a sounding board) distinct raps will be heard. The effect is most deceptive, because the movements producing the "impulses" are so minute as to pass unnoticed.

You can next do a "quickie" effect with the stick, causing it to appear to cling to the palm of your hand when it is turned downward. This is the trick you have already learned in which you grip your right wrist with your left hand, and the left forefinger secretly supports the stick on its underside (see page 16). You can have a spectator reach out and pull the stick away from the palm, folding your left forefinger to join the others on the wrist as this is done. Performed entirely as an incidental, it adds a little extra touch to the routine.

4. The Magnetized Playing Cards

For this spectacular demonstration in the routine, you apparently magnetize half a deck of cards so that they cling haphazardly to the fingers and palm of your hand. Actually, the effect depends on a finger ring that you will wear for your performance on your middle finger.

To accomplish this effect, a small strip of transparent plastic (acetate) of about twelve gauge is used. The strip should be about one-quarter inch wide and two inches long. The ends of this little strip of clear plastic should be curved. Sand the curves down until smooth. To perform the trick, secretly slip this strip under your finger ring on the inside of the hand (if you are right-handed, have ring on your left hand or vice versa if you are left-handed), so that the strip goes halfway under the ring. Half of the strip will then protrude into the palm and the other half will lie along the finger, as shown in the drawing. Since the plastic is transparent, the palm may be flashed as long as the hand is kept in motion, and the strip will not show.

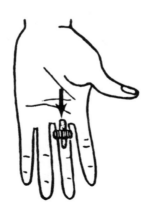

When you are ready for the performance, place your palm flat on the table. Take two cards and slide them under your hand, so that they go between the plastic strip and the palm. Now slide two more cards underneath from the opposite sides of the hand so they also lie under the transparent strip. You now have four cards under the strip, facing in four directions. Other cards are now slipped underneath the hand haphazardly, but so that they go between the hand and any of the first four cards. In this way, you can distribute half a deck of cards so that they overlap the edges of your hand at all angles.

Proceed to raise your hand from the table, and all the cards are seen to cling to the hand as if magnetized. You can turn your hand in all directions as the clear plastic strip cannot be seen against the cards while you swing them about.

When you are ready to release the cards, turn your hand downward and close the hand by bending the fingers slightly. The original four cards held by the strip will glide off, and all the cards will flutter down to the table. While the cards are being examined, you'll have ample time to rid yourself of the small gimmick beneath your ring.

5. Magnetic Table Lifting

The climax of the whole routine too is accomplished by your finger ring. The table used is a lightweight one, which you can use throughout your performance. The top of this table is painted black and a small nail (the type with a head) is driven into its surface near the center so only a tiny bit of it projects from the tabletop. The projecting nail head is also painted black, and is invisible. The drawings on page 28 show both the effect of the trick and this secret preparation.

Your finger ring, which you again wear on your middle (medius) finger, has a slot sawed halfway through it, as shown in the drawing on page 28.

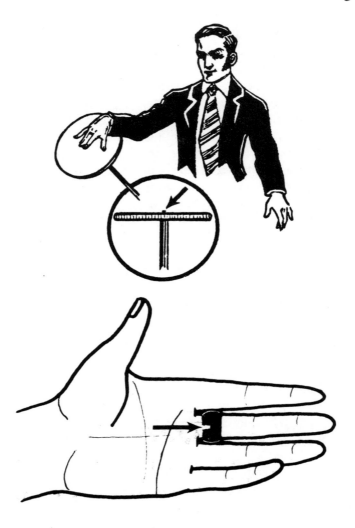

When you are ready to perform the "magnetic" lifting of the table, remove all articles from the table and show it casually; also show your hands. Then rub them briskly together to appear to build up the magnetism, then place both your hands side by side on the surface of the table. The hand wearing the slotted ring goes directly over the tiny projecting nail head; you engage this in the slot of the ring. By exerting a little pressure on the surface of the table with your fingertips, you will hold the table rigidly to your hand, and when you lift your hands up into the air, it will cling to them as though by magnetism.

Having shown the table floating from both hands, remove your free hand and wave it about the table, thereby showing indirectly that there are no connections between hand and table. The table is then floated back to the floor, and your other hand is freed.

You can again show as though indirectly that there is no connection between your hand and the surface of the table by picking up a thin handkerchief or scarf and draping it over the top of the table. Again place your hand on the table, right on top of the scarf, and slip the ring notch over the projecting nail head. The scarf does not interfere with your engaging the nail head in the slot of the ring, and you can once more lift the table "magnetically." The drape perfectly conceals the modus operandi, and you can walk among the spectators with the table clinging to the palm of your hand. It is very effective.

To conclude the trick, return with "The Magnetized Table" to the front, disengage the ring, and your Magical Magnetism Act is complete.

2

MAGIC WITH NUMBERS

It is important for everyone to know arithmetic, since it is used throughout life, yet many people find it awfully dry. Numbers can be so factual as to seem uninteresting. But by applying magic to numbers you can bring them to life and make them sparkle with mystery.

Arithmetic can lead on into the science of mathematics, which expresses nature in terms of symbols and is therefore truly magical.

Tricky Numbers

Numbers are wonderful and mighty tricky. Here are some fun things to know:

When two even numbers are added together or subtracted from each other, their sum or difference will always be an even number.

When two uneven numbers (or "odd" numbers) are added together or subtracted from each other, their sum or difference will likewise always be an even number.

The sum or difference of an odd and even number will always be an odd number, while the sum of two odd numbers will always be an even number.

If two different numbers may be divided by any one number, their sum or difference may also be divided by that number.

If several different numbers that may be divided by 3 are added or multiplied together, their sum and their product may also be divided by 3.

If two numbers that may be divided by 9 are added together, the total will be either 9 or a number divisible by 9.

When any number is multiplied by 9, or by any other number which is divisible by 9, the product will be either 9, or a number which may be divided by 9.

To guess the number a person is thinking of, use this trick with numbers. To do this, ask a spectator to think of any number from 1 to 9. Let's suppose that the person thought of the number 6. Ask him or her to multiply the number thought of by 3 (which for our example will make a product of 18.) Tell him or her to add 1 to this product (which, in this case, makes a total so far of 19). Have your spectator multiply that total by 3 (it comes out here to 57), then add to this product (57) the number secretly thought of. For our example, the final total is 63. Have the person then tell you the final number produced; it will always end with 3. Strike off the 3, and inform the spectator that he or she thought of the number 6. Try it with different numbers (from 1 to 9). It works every time and seems like mind reading.

Add a Number

The Effect:

This is a fast trick with numbers that is so clever you can almost fool yourself as you learn how to do it. For the trick, have a person think of any number between 1 and 10. Tell them to double the number of which they are thinking, then add 10 (or any other even number, but you tell them which) to the doubled number. Now have them divide the total by 2 and subtract the original number. Your answer is "Five!" You tell 'em every time.

Here's How You Do It:

The trick works itself automatically. For example, let's suppose that the number the person thought of was 8. Then 8 is doubled, making 16. Next, 10 added to 16 equals 26, and 26 divided by 2 equals 13. Subtracting the original number thought of (8 in this instance) from 13 leaves 5. *The result will always be one-half of the number they are told to add.* If 10 is added, all answers will be 5. So, when you perform the trick, vary the number you tell them to add to their doubled number so as to get different results and mystify the spectators.

The Age and Loose Change Trick

The Effect:

Hand a person a piece of paper and a pencil for them to figure this one on. Ask the person to think of his or her age and write it down on the paper, without showing it to you.

Now tell the person to multiply the number by 2, then add 5 to this answer and multiply the resulting number by 50. From this grand total, have your subject subtract the number of days there are in a year (365). Now have him or her count up the amount of small change he or she is carrying in pocket or purse that is under one dollar; and next to add this amount in numbers (66 cents being the number 66, for example) to the total so far. Ask the person participating in the stunt to give you the final answer. Using this figure, you are able to tell his or her age and the amount of change (under a dollar) he has in his pocket.

Here's How You Do It:

So you will understand exactly how this operates, let's take this example: suppose a man's age is 35. In this instance, thus, 35 multiplied by 2 equals 70. When 5 is added to this, it comes out to 75. 75 is now multiplied by 50 and the grand total is 3,750. Subtracting 365 from this amount makes 3,385. Now, let's say the small change the man has in his pocket under a dollar is 50 cents, so 50 is added to 3,385, and the total comes to 3,435. The person tells you this total.

As soon as you know the total (in this example, 3,435) you mentally add 115, and your mentally figured total comes to 3,550. And there you have it: The first two numbers of that figure give you the person's age (35), and the second two numbers give you the amount of change he has in his pocket (50 cents).

It's a great stunt with numbers to know, and works with any age and any amount of money under a dollar.

Family-History Arithmetic

The Effect:

Give a spectator a sheet of paper to figure on while you turn your back so that all will be kept confidential. First, ask the spectator to put down the number of brothers that he or she has. Then have that number multiplied by 2; add 3 to this total, and then multiply the total so far by 5. To this answer, tell the spectator to add the number of sisters that he or she has, and to multiply this result by 10. Then, for good measure, ask the spectator to add his or her "lucky" number (that is, any number from 1 to 10) to this final figure. Now, request the spectator to tell you the result of this figuring. You can tell how many brothers and how many sisters he or she has and what the "lucky" number is. Very revealing arithmetic indeed!

Here's How You Do It:

When the person gives you the final result, you subtract 150 from this number. The first figure of the answer tells you the number of brothers, the second figure the number of sisters, and the third figure reveals the lucky number. Look at the example that follows and you will see exactly how to do it every time:

Let's say, in this case, that the spectator has three brothers; 3 multiplied by 2 equals 6. He or she adds 3 to this (6) and gets 9; then multiplies 9 by 5 and gets 45. Next the number of sisters is added in, let's suppose it is two, so 2 added to 45 makes 47. Next this result is multiplied by 10, which comes to 470. To this the spectator adds a lucky number. Let's say the spectator liked the number 5: The grand total is, then, 475. He or she tells you this final figure. From the grand total, you mentally subtract 150 and you arrive at the number 325.

325: thus, you learn that the spectator has three brothers and 2 sisters, and 5 is the lucky number he or she favors. It works every time, and you can learn the history of anyone in this mathemagical manner.

Funny Arithmetic

Here are two arithmetic novelties from the *Magic Souvenirs* of the outstanding woman magician Dell O'Dell, as illustrated by the famous cartoonist-magician H.C. Bjorklund. They are fun....

Here is a little problem in adding (not like school adding).

Cross out some of the numbers on the slate so that those remaining will add up to 1,111.

After you have solved it, try it on your friends and see how many of them can work it out.

SHHHHH! HERE'S
THE ANSWER

$$
\begin{array}{r}
111 \\
3 \\
7 \\
99 \\
\hline
1111
\end{array}
$$

A Sum in Addition

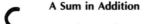

Strike out six figures shown in the sum so that those remaining will add up to 20.

To get 20, strike out the first 1, all the 7's, and the first two 9's; to get 100, strike out the first two 1's, all the 7's, and the first 9.

Guessing the Erased Number

The Effect:

Allow yourself to be blindfolded and ask a spectator to write down a four-digit number such as 8,930, 6,257, 1,974; any four-digit number that he desires. In this example, let's say the number secretly written was 4,872. You then ask that he multiply this number by 9. The result comes out to be 43,848.

Then, instruct the spectator to select any digit of this large number except zero and erase it (if the spectator is working on a slate or blackboard he erases it; if on a pad he crosses it out). Then claim that you will be able to tell him the number he has erased if he gives you just one bit of information—the sum of the remaining digits. In this example, if the digit 3 was crossed out, the sum of the remaining digits (4,8,4,8) would be 24. As soon as he tells you this number, you immediately inform him that the number he erased was 3!

Here's How You Do It:

To find which number was erased or struck out, you subtract the sum of the remaining digits from the nearest multiple of 9 which is higher than the sum in question. In this example, the sum being 24, the required multiple is 27. Subtracting 24 from 27 gives you 3. You have correctly divined the erased number.

The Five Odd Figures

The Effect:

Announce that you can write five odd figures that will add up to an even number— 20. Let the spectators try to do it, and they will agree that it is impossible because an odd number of odd numbers always adds up to an odd total. But you do it!

Here's How You Do It:

You have cleverly said five *figures*, not numbers, so for the first figure you write down 1, then place beside it the second figure you write down, 7—this gives 17. Then for the third figure you write under the 7 a 1, then for the fourth figure another 1, and for the fifth figure another 1, as shown...

$$
\begin{array}{l}
17 \\
1 \\
1 \\
\underline{1} \\
20
\end{array}
$$

17 You have written down
1 five odd figures which
1 on being added together
1 total an even number.
20

Draw a line under the five figures you have written, and you get the even number 20 as the total. You did it!

Which Hand Holds Which Coin?

The Effect:

This is number magic using two coins. While your back is turned, the spectator hides a dime in one hand and a penny in the other. By a simple calculation you know which hand contains which coin, and can inform the spectator accordingly.

Here's How You Do It:

The spectator is asked to multiply to himself the value of the coin in his right hand by 4, 6, or 8, and the value of the coin in his left hand by 3, 5, or 7. Tell him to add the results and give you the answer.

If the answer is an even number, the penny is concealed in his right hand. If it is odd, then the penny is concealed in his left hand and the dime in the other.

The Psychic Book Test

This is a "big-time" trick from the programs of famous mentalists. "Mentalist" is the title given to magicians who feature mental magic. This kind of conjuring is of great interest to many people, and some consider it the most "grown-up" of all forms of magic. The present mystery gives you a chance to combine number magic with mind-reading magic.

The Effect:

Hold up a large book and explain to your audience that it contains thousands of words, and that you are going to attempt an experiment to locate the one word out of all those thousands that is selected at random. In order to make the selection absolutely fair, you state that it will be handled by mathematics.

At this point in the effect, you give the book to a spectator to hold and examine, and then you pick up a pad and pencil. Three different members of the audience are asked to call out any three single-digit numbers, such as 5, 8, 1 for example. They may call out any numbers they wish. As the numbers are called you write them on the pad. The pad with the three numbers written on it (581 in this instance) is given to a spectator who is requested to reverse the numbers, to obtain in this case 185. He is now asked to subtract the smaller number from the larger. You ask that he call out the result. In this instance, it is 396. This number, as called, you write on another pad of paper, and hand it to another spectator. This spectator is then asked to reverse the number and place it under the original number. You then ask that these numbers be added together. The total comes out as 1,089. You can follow this readily by this outline:

EXAMPLE:		
Three numbers called were		581
These numbers are reversed under above		185
Subtract		396
396 is then written on second pad and handed to another spectator		396
396 is then reversed under above, making		693
Numbers are added together, totaling		1,089

You then explain that this number determined in this random manner will be used in two parts to determine which word will be selected in the book—the first three digits of same (108) being used to locate the page in the book for the test, and the last digit of the number (9) to locate the word on that selected page.

Accordingly, you ask the person who has the book to turn to page 108, and on that page to count down nine lines from the top, and concentrate on the first word of that line. The spectator does as directed. Dramatically, you "read his mind" and call out the very word upon which he is concentrating. Or, if you prefer, you can hand the spectator a slate and a piece of chalk, and request that he write his selected word on it boldly while your back is turned. As he does this, while standing at a distance, you write upon another slate a word. When the slates are compared, the words match.

Here's How You Do It:

This remarkable feat of mental magic is accomplished by what is known as a "force." In other words, you know in advance what word is going to be selected. You have memorized this word (the word in the book you are using that is the first word of the ninth line on page 108). The ingenious handling of the numbers (as described in the effect) will always come out to 1,089 no matter what the original numbers called, so while it seems all is fair and aboveboard, actually you have *forced* the desired word in this arithmetical manner unbeknown to the spectators. When you ultimately reveal the word upon which the spectator is concentrating, the effect is baffling.

The 1,089 Card Trick

You can use this 1,089 principle to make an interesting card trick. Lay out a ten-spot card, an eight and a nine face down in a row on the table. Have a spectator call out any three-digit number, then reverse the digits and subtract the smaller number from the larger. By reversing the result and adding these numbers together, you will always arrive at the final result of 1,089, as you have learned. All that remains to be done is to dramatically turn over the cards on the table: first the ten, then the eight and finally the nine-spot card...and you have 10 8 9 (1,089).

The Human Calculator

The Effect:

Five rows of numbers are written. Any person may write the first, second, and fourth rows. You write the third and fifth rows. You draw a line beneath the figures, and immediately add all the numbers at a glance—correctly writing the total below the line.

Here's How You Do It:

The secret is in writing the third row of numbers. As you do this, you take care to bring each figure of the second row up to a total of 9. In other words, under a 6 you write 3; beneath a 1 you write 8, etc. In the same way, you write in the fifth row so that it brings the figures of the fourth row up to 9. To get the result in this exercise in lightning calculation, simply write down the top horizontal row of figures less 2 taken from the last number of the group, and put the 2 in front. The columns here illustrated show exactly how it is done:

$$
\begin{array}{ll}
\quad\ 2 & \\
38427 & \text{spectator's figures} \\
40314 & \text{spectator's figures} \\
59685 & \text{your figures} \\
28087 & \text{spectator's figures} \\
\underline{71912} & \text{your figures} \\
238425 & \text{THE INSTANT TOTAL}
\end{array}
$$

How to Be a Lightning Calculator

The Effect:

You offer to demonstrate your ability as a "lightning calculator." First, say that you will multiply any three-digit number given you by 143, and do it all in your head quickly. A little figuring, and you call out the correct answer. The spectators will have to figure it out the long way on a piece of paper. You are correct. For the second part of the effect, you ask members of the audience to select any number from 1 through 100 and you are able to give the cube root extraction of it right from the top of your head. These tricks make you look like a mathematical genius.

Here's How You Do It:

For the lightning multiplication of any three-digit number by 143, follow the procedure shown below with 587 as an example.

Take the number 587 and picture it in your mind as 587587 and divide it by 7. Dividing the number in this manner mentally is not difficult with a little practice. Your answer to the multiplication of these two numbers by this tricky method comes out 83,941. You are correct; let the spectators multiply the numbers together in their long way, and they'll think you're a whiz at math. This "lightning calculation" trick will work with any three-digit number given you.

Here's another instance: Let's multiply 307 by 143. Picture in your mind 307 as 307307 and mentally divide by 7: 7 goes into 30 four times and 2 over. 7 goes into 27 three times and 6 over. 7 goes into 63 nine times. Now, 7 times zero (0) remains zero (0) and 7 into 7 is one, so the answer is 43,901.

The cube root extraction part of the demonstration is accomplished in this manner. Have a member of the audience select any number from 1 through 100, cube it on a pad of paper, and then call out the result. Figuring it out in your head, you give the cube root extraction (the basic number that was cubed) correctly. To do this, first you must have memorized the cubes of numbers from 1 through 10:

$$
\begin{array}{rl}
1— & 1 \\
2— & 8 \\
3— & 27 \\
4— & 64 \\
5— & 125 \\
6— & 216 \\
7— & 343 \\
8— & 512 \\
9— & 729 \\
10— & 1,000
\end{array}
$$

An inspection of this table reveals that each cube ends in a different digit. The digit corresponds to the cube root in all

cases except 2 and 3, and 7 and 8. In these four instances, the
final digit of the cube is the difference between the cube root
and 10.

By way of an example, let's suppose that a spectator asks
you what number is cubed to make 250,047. The last number
of this figure is 7, which tells you immediately that the last
figure of the basic number cubed is 3. And the first figure of
the basic number cubed is determined by discarding the last
three figures of the cube (regardless of the size of the numbers)
and retaining the remaining figures, which in this case are 250.
In the above table you have memorized, 250 lies between the
cubes of 6 and 7. The lower of the two figures—in this instance, 6—is the first figure of the basic number that was
cubed. The correct answer of this cube root extraction is therefore 63.

One more example will make this handling clear. Say the
number 19,683 is called out. The last digit, 3, indicates that
the last digit of the cube root is 7. Discarding the final three
digits leaves 19, which falls between the cubes of 2 and 3. 2
is the lower number, therefore you arrive at the final cube root
of 27. It's an impressive trick that you can perform if you have
a leaning toward math.

I Have Your Number

The Effect:

On a card, write the number 142,857 and ask a spectator
to multiply it by any number between 1 and 6. You claim that
you already have the answer no matter what number the spectator selects to multiply by. To prove this, you bring out a
strip of paper on which a line of numbers is written, tear this
strip into three pieces, and give these to three spectators. Then

ask the first spectator what was the product of his or her multiplication (as an example, 4 × 142,857 = 571428). You then collect the strips of paper with the numbers on them, place them together in a line on your table, and show that you knew what the answer would be in advance.

Here's How You Do It:

Number 142,857 is a special number that, when multiplied by any number between 1 and 6, will give a product result which will always contain the numbers 1, 4, 2, 8, 5, 7 in a cyclic order. Thus, you already have 1 4 2 8 5 7 written on the strip of paper, and when the spectator calls out his number, you have only to tear the strip into three pieces according to the number (from 1 through 6) chosen, and reassemble them as shown in the drawing. By the way in which you line up the three pieces of paper you can show that your advance prediction of the total was correct.

HE ONLY TEARS OFF THE STRIP
ACCORDING TO THE NUMBER
SELECTED BY THE SPECTATORS.
142857 × 1, 2, 3, 4, 5, 6 respectively-

1 4	2 8	5 7
2 8	5 7	1 4
4 2	8 5	7 1
5 7	1 4	2 8
7 1	4 2	8 5
8 5	7 1	4 2

Some Number Funnies

Write down this column of numbers:

$$
\begin{array}{r}
98 \\
87 \\
69 \\
49 \\
22 \\
54 \\
67 \\
76 \\
\underline{83} \\
605
\end{array}
$$

Add them all together and you get a total of 605. Now
completely reverse the numbers and form another column so
98 becomes 89, 87 becomes 78, and so on, as shown:

```
        89
        78
        96
        94
        22
        45
        76
        67
        38
       ───
       605
```

Now add these numbers and you get the same total as in the first column before the numbers were reversed.

Can you subtract 45 from 45 and still have 45? You can if you do it this way:

$$
\begin{array}{rl}
9\ 8\ 7\ 6\ 5\ 4\ 3\ 2\ 1 & = 45 \\
-1\ 2\ 3\ 4\ 5\ 6\ 7\ 8\ 9 & = 45 \\
\hline
6\ 6\ 4\ 1\ 9\ 7\ 5\ 3\ 2 & = 45
\end{array}
$$

Here's a new way to multiply 69 by 47:

To get the product result, divide the figures in the left-hand column by 2 and double the figures in the right-hand column.

Now draw a line through all the even numbers in the left-hand column and all the numbers opposite them in the right-hand column, as here shown:

DIVIDE	DOUBLE	ADD
69	47	47
~~34~~	~~94~~	
17	188	188
~~8~~	~~376~~	
~~4~~	~~752~~	
~~2~~	~~1504~~	
1	3008	3008
		3243

Then take the numbers in the right-hand column that have not been eliminated and add them for the result. And you'll be right!

This may be doing things the long way around, but the stunt is a lot of fun if you perform the figuring on a blackboard, so everyone can see how you worked it out.

Ask a spectator to give you his or her favorite number from 1 to 10. Let's suppose the chosen number is 6. Write down all the digits as follows, skipping the 8: 1 2 3 4 5 6 7 9. Multiply the favorite number by 9 which, in this instance, gives 54. Then multiply the row of figures from 1 to 9 inclusive (leaving out the figure 8), by 54 and you will come up with a long row of 6's (666666666). It works with any number, so you always produce the favorites.

To perform the trick, simply multiply whatever number is selected by 9 every time. Then multiply the row of figures by the result of the favorite number multiplied by 9, and it will always give you a long row of the favorite number. Here's an example of how it works for number 5:

Multiplying 5 by 9 gives 45, and 12345679 multiplied by 45 gives nine fives, as...

$$\begin{array}{r} 12345679 \\ 45 \\ \hline 61728395 \\ 49382716 \\ \hline 555555555 \end{array}$$

A Number Game

The Effect:

This is a game played with numbers that is good practice in arithmetic at the same time. You and a spectator play the game together: The object is a number race to see who will reach 100 first. Taking turns, you each call out numbers jumping by 10 or less each turn. As a whiz with numbers, you win the game every time. Then on the QT you show the spectator how to play it always to win also. Then he or she can join in and have some fun playing it with others.

Here's How You Do It:

You win in this game if you manage to reach 89 first, for your competitor cannot then add more than 10 or less than 1. So, on your next turn, the winning score of 100 is yours for the taking.

To reach 89 first, simply capture the "key numbers" of 45, 56, 67, and 78. These numbers are easy to remember because, like 89, the second digit is always one more than the first digit. Once you reach a "key number," from then on whatever number your competitor adds you simply add to it a number that will make 11, *and you'll always be a winner*.

It's a fast game of mental addition, and after you have handily beaten the spectator, show him or her how to do it to be a winner too.

The Mysterious Number Nine

Nine (9) is a number of cabalistic power, the trinity of trinities according to Christians, the number of perfection according to the Greeks, and the superlative of superlatives according to the Hindus.

Nine is a mysterious number: Different beliefs teach that there are nine heavens, nine orders of angels, nine planets that influence mankind, nine lives of a cat, nine days of mortification, nine regions of hell, nine heads of the hydra; modern leases last for 99 years, ancient leases for 999 years; the cat-o-nine tails suggests punishment and atonement; the figure 9 upside down is 6, and 666 is the number of the Antichrist. The ancients knew nine as the irrepressible number. Whenever it is used as a factor in a mathematical calculation, it is bound to come out in the result.

The figure 9 is the last of the digits, and the highest number that can be expressed in one digit. And you can arrange four nines (9's) so they equal 100.

$$99\frac{9}{9}$$

3

MAGIC WITH CHEMISTRY

There was a time in the distant past when magic and chemistry were combined in one art. This was medieval alchemy, a supposed science that sought magical results by chemical methods such as changing base metals into gold. It is believed that alchemy had its origin among the Greeks of Alexandria, was adopted by the Arabs, and was then handed over to the Europeans. In Europe, it developed into the science of chemistry.

Although alchemy failed in many of its magical aims, its contributions to modern chemistry are numerous, both in basic laboratory apparatus and in chemical combinations. Modern chemistry might still be called magic, considering the "miracles" scientists perform, such as creating plastics, dyes, medicines, and any number of other useful things. Like medieval alchemists, today's scientists earnestly desire to learn the secrets of nature. The world is full of chemical wonders yet unexplored.

WARNING: CHEMICALS CAN BE DANGEROUS. HANDLE THEM WITH CARE AND KEEP THEM OUT OF THE REACH OF CHILDREN AT ALL TIMES.

Magic with chemistry is an amazing way to stimulate interest in science and to teach about its operations. Chemistry and magic go together—each complements the other. The reactions of chemicals are truly like magic. You will find less explanation of the various effects in this chapter than in the others in this book: The chemical reactions simply occur as described.

The chemicals used in the experiments that follow can be obtained easily through drugstores or chemical supply houses. *Chemicals must be handled very carefully, and they must be kept out of the reach of children at all times—including during your performances.* It is wise to caution the audience not to touch your apparatus when you present chemical magic. This has the double effect of safeguarding the audience and arousing their interest and curiosity.

A NOTE ABOUT CHEMICAL SOLUTIONS: Be sure to shake chemical solutions well before using them. You can experiment with them a little to get the exact proportions of dissolved chemical (forming the solution) that work best for you. Keep your solutions stored in stopped glass containers (bottles) and label them so you will know the contents at a glance. The experiments in this book use relatively small quantities of chemicals, and so the solutions are usually clear. Therefore, you must exercise particular care to keep them separate and straight in your mind. Never combine chemicals at random. Follow the instructions you are given as closely as possible.

The Wine and Water Trick

You will need:
  a glass pitcher
  5 tumbler glasses
  phenolphthalein
  powdered *tartaric acid potas-
    sium carbonate* or *sodium
    carbonate*
  water

The Effect:

A glass pitcher half filled with water is set out, along with
five empty glasses. The performer pours water into the first
glass, wine into the second, water into the third, and wine
into the fourth. Then he pours the contents of the first and
second glass back and forth, and they both become wine. He

pours the contents of the third and fourth glasses back and forth, and they both become water. The first and second glasses of wine are poured back into the pitcher, making wine in the pitcher. The third and fourth glasses of water are poured back into the pitcher, and the wine in the pitcher visibly changes to water, as it was at the beginning of the trick. For a surprise climax, the magician then pours water from the pitcher into the fifth glass, and it changes to milk!

Here's How You Do It:

From a drugstore, obtain a strong solution of phenolphthalein. (This is made from filling a bottle about one-fifth full of powdered phenolphthalein and then filling the bottle to the top with alcohol.)

Fill a second bottle half-full with powdered tartaric acid, then fill the bottle up to the top with water.

Fill a third bottle half-full with potassium carbonate or sodium carbonate, then fill the bottle up to the top with water.

The chemicals for performing this trick are now ready. You also need a clear glass pitcher with enough water in it to fill four glasses, and five glass tumblers.

To prepare for the demonstration, arrange the glasses in a row on the table, and into the first glass, place ½ teaspoon of the potassium carbonate solution; into the second glass, place a few drops of the phenolphthalein solution; into the third glass place 1 teaspoon of the tartaric acid solution; into the fourth glass, place a few drops of the phenolphthalein solution, and into the fifth glass, place 1 teaspoon of the phenolphthalein solution.

You are now ready to work the trick. As was outlined in the Effect section, you pour water from the pitcher into the first glass, filling it about two-thirds full. The water mixes with the clear solution of potassium carbonate in the glass and looks like water. Pour the contents of this glass back into the water pitcher. This action will make a mild solution of potassium carbonate in the pitcher.

Again you pour the "water" from the pitcher into the first glass, and place it on the table as a glass of water.

Then, pick up the second glass, and pour the "water" from the pitcher into it; the chemical reaction occurs, and the glass appears to be filled with wine.

Pick up the third glass and pour the "water" from the pitcher into it; it mixes with the tartaric acid but no visible reaction occurs, so it passes for a glass of water.

Pick up the fourth glass and pour the "water" from the pitcher into it. It will appear as though you have poured another glass of wine. The visible results on your table will be two glasses of water and two of wine that have been poured from the pitcher of water alternately.

Now comes more magic. Pick up glasses No. 1 and 2 and pour the contents back and forth between them. The result appears as two glasses of wine. The "wine" is poured from the glasses into the pitcher, which results in the entire contents of the pitcher turning red. You show it to the audience as a pitcher of wine.

Now pick up glasses No. 3 and 4 and pour the contents back and forth between them. The result is two glasses of "water." The "water" is poured from the glasses into the pitcher, and the chemical reaction here changes the "wine" in the pitcher into "water." So you end up with what appears to be a pitcher of water, the same as when you started the trick.

For the climax of the effect, pour the "water" from the pitcher into glass No. 5, and you end up with a glass of "milk"[1].

---

[1]See the paragraph about chemical solutions on page 52. These solutions are clear, and the relatively small quantity that goes into each glass is not noticed by the spectators who may be seated only a short distance from your table.

If you wish to add an additional comic touch to the trick, you can have a sixth glass on your table in the bottom of which is a little liquid soap and iodine. When you pour the "water" from the pitcher into this mixture, you appear to produce a glass of beer.

This trick is based upon the following chemical reactions: Phenolphthalein turns the color of red wine when an alkaline solution enters it. An acid solution will turn it colorless again, making it appear as water. Thus, in this chemical magical feat, potassium or sodium carbonate, being alkaline, turns the phenolphthalein solution red; the tartaric acid, being acid, bleaches it out again, and produces the water effect. The milk effect, in the fifth glass for the climax of the trick, is produced by pouring water into the strong phenolphthalein solution.

Is It Wine Or Is It Water?

You will need:
  a glass pitcher
  a glass tumbler
  sodium hydroxide solution
  phenolphthalein solution
  sulphuric acid

This is another "Wine and Water Trick," but this one carries a double color change, which comes as a complete surprise to the audience.

The Effect:

Water is poured from a pitcher into a glass and it turns to wine. The wine is held up before the spectators. What stuns the audience is that even as they are looking at the wine, it suddenly changes back to water.

Here's How You Do It:

Place a small amount of concentrated sodium hydroxide solution in the glass. It is clear, and the quantity in the glass is so small it will not be noticed. (Sodium hydroxide can be purchased as a solution from a chemical supply house.)

Fill the pitcher with water, some phenolphthalein solution, and enough sulphuric acid so that one glassful of the solution will just neutralize the sodium hydroxide solution in the tumbler. You can experiment in private to get just the right proportions.

In performing the trick, when you pour this "water" into the glass it changes to a bright wine color. Look at it, smack your lips, and call it wine. Tell your audience to concentrate their attention on the wine, for right before their eyes magic is going to occur. While everyone is watching closely, suddenly the "wine" will change back to "water."

This kind of a delayed change in chemistry is known as a "clock reaction." The following is another chemical trick based on this principle.

Capturing Smoke in a Glass of Water

You will need:
  a glass pitcher
  a glass tumbler
  a handkerchief
  a piece of paper
  sodium thiosulphate
  water
  sulphuric acid

The Effect:

For this trick, you pour water into a glass and cover it with a handkerchief. Burn a rolled-up piece of paper and fan the smoke in the direction of the glass. When the handkerchief is removed, the smoke will seem to have gone into the water, which now presents a milky, clouded appearance.

Here's How You Do It:

The "water" used in the trick is actually a little sodium thiosulphate (known as "hypo" and obtained from photographic supply stores) dissolved in about a pint of water. Put a few drops of sulphuric acid in the glass. Pour some of the diluted hypo solution into this tumbler. Because the chemical reaction takes a while to occur, you have ample time to cover the glass. Then go through the business of blowing the smoke toward the glass. On removing the handkerchief, the liquid will seem to have captured the smoke.

The Eggs that Read

You will need:
  3 jars of quart size or larger
  water
  salt brine (*sodium chloride*)
  a funnel
  hydrochloric acid
  3 eggs
  a dark-colored crayon

The Effect:

Set three jars (any sizes) filled with water and a plate of eggs on your table. Announce that these are very special eggs as they have the ability to read. Pass the plate to someone in the audience and have them select three eggs. Using a crayon, on one egg write the word *sink*. On the second, write the word *float*, and on the third, write the word *swim*.

Then the magic starts: When you drop the egg with the word *sink* in one of the jars of water it sinks to the bottom. The egg marked *float* sinks and then floats midway in the second jar of water. The *swim* egg, on being dropped into the remaining jar of water, swims about and rises to the top and falls to the bottom several times.

Here's How You Do It:

Fill one quart jar with clear water. Fill the second jar half-full with salt brine (saturated solution of sodium chloride). Then, by using a funnel and pouring the water in slowly so as not to disturb the brine in the lower half of the jar, fill the

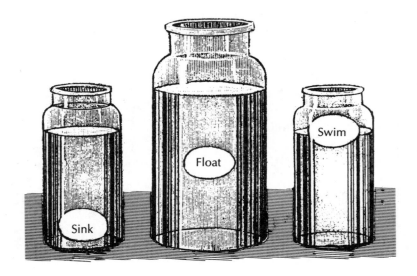

jar to the top with clear water. Fill the third jar with a diluted
solution of hydrochloric acid. All jars appear to be simply filled
with water, yet the following surprising results will be attained:
When the egg with SINK written on it is dropped into the
jar of plain water, it will sink to the bottom. When the egg
labeled FLOAT is dropped into the jar that is half salt water,
it will fall through the water and will stop to float midway in
the jar. (The principle of density of solutions is demonstrated
here.) When the egg labeled SWIM is dropped into the jar
containing the diluted acid, it will rise and fall and swim about
in a mysterious manner.

The Amazing Finger Ring

You will need:
a finger ring
thread
matches
alum

The Effect:

A finger ring, borrowed from a member of the audience, is tied to the end of a length of thread. The thread is then tied to a suitable support with the ring dangling down on its end. Taking a match, you set fire to the middle of the thread, which burns from one end to the other—yet the ring does not fall.

Here's How You Do It:

Soak the thread in a strong solution made by dissolving alum in water. Remove the thread and let it dry. Then soak it in the alum solution again. Repeat this process several times. Now suspend the ring as described in the Effect section, and set fire to the thread in its middle. Amazingly, the ring will not fall, for although the thread is really burned, the alum forms a tube of chemical substance that has sufficient strength to support the ring as it dangles in the air.

The Magic Flower Seeds

You will need:
  copper sulphate
  iron sulphate
  cobalt chloride
  manganese sulphate
  plaster of paris
  water
  sodium silicate solution

The Effect:

Bring out a small pillbox in which you say you keep your "magic flower seeds." Displaying a bowl of water, you offer to show the spectators a demonstration of how these wonderful seeds will grow into a beautiful flower garden. When you drop a seed in the water, it begins to sprout, producing a most enchanting floral display.

Here's How You Do It:

This is real chemical magic. To make the "magic flower seeds" sprout and blossom, take twelve drams (1½ oz)[2] of copper sulphate, two drams (¼ oz) of iron sulphate, two drams (¼ oz) of cobalt chloride, twelve drams (1½ oz) of manganese sulphate, eight drams (1 oz) of moist plaster of paris (powder with water added) and mix with enough water to make a thick paste.

From the paste formed of the above combination, mold

[2]1 fluid dram = ⅛ fluid ounce.

your "magic flower seeds" to about the size of lima beans. These will harden, and you can carry a supply of them in a pillbox, ready for a performance at any time. The "water" in the clear glass bowl in which you drop the "seeds" is a solution formed of a quarter of a pint of "water glass" (sodium silicate solution) and a half pint of plain water.

The chemical reactions that occur when you drop your "magic flower seeds" into this solution cause a chemical "flower garden" to grow in a short time.

The Color Changing Powder

You will need:
   a small cardboard box with a
      lid
   potassium iodide (powdered)
   lead nitrate (powdered)

The Effect:

You show a little cardboard box filled with white powder. Give the box to a spectator and request that he shake it while he says what the powder is. Naturally he says "white." But when the box is opened, the powder has turned to brilliant yellow. It's magic!

Here's How You Do It:

The box contains finely powdered *potassium iodide* and powdered *lead nitrate*. Both of these are white in color and are placed in the box separately, one over the other. They remain white while at rest, but when mixed together by the shaking the spectator gives the box, they turn yellow.

Chemical Novelties

Here are a dozen unusual demonstrations using chemicals. Each of the stunts carries a magical touch:

1. The Indestructible Card

You will need:
tinfoil
a calling card

You won't believe this, but you can do it. Place a small (one inch by one inch) piece of tinfoil on a calling card, and hold the card over the flame of a candle. The tinfoil will melt, but the card will not burn.

2. The Restless Mothball

You will need:
  acetic acid
  sodium carbonate
  a test tube
  a mothball

Arrange a large test tube so it stands upright. Place enough acetic acid in it to fill the tube almost to the top, then add a little sodium carbonate, which will generate a gas. When you drop a mothball into this solution, it will become very rest-less—it will rise and fall several times within the test tube. These interesting results are caused by the gas forming on the mothball, which is expelled upon reaching the surface.

3. The Darting Camphor

You will need:
  camphor
  a shallow basin
  water
  matches

This makes a good follow-up stunt to "The Restless Moth-ball," since you cause a piece of camphor to become even more restless than you did the mothball. To perform this feat, place a piece of camphor in a shallow basin of water. Light the

camphor, and it will dart about on the surface of the water
like a miniature jet.

4. Changing Two Liquids
into a Solid

> You will need:
>   *calcium chloride*
>   *potassium carbonate*
>   hot water

Ask your audience if they believe that you can pour two
liquids together to produce a solid. This usually arouses interesting speculation. You *can* do it. Make a saturated solution
of calcium chloride dissolved in hot water. (Saturation is the
point at which no more calcium chloride will dissolve.) Set it
aside to cool. Prepare a solution of potassium carbonate in the
same manner. Use Pyrex (heat-resisting) glass containers in
preparing the solutions. They appear as two liquids. When
you pour these cooled liquids together, a solid will be produced.

5. The Tricky Soap

> You will need:
>   aniline dye
>   soap
>   a basin
>   water

Soap is supposed to clean one's hands; this soap does just the opposite. The stunt is easily prepared and provides much laughter. What you do is to rub some aniline dye—any color—on a cake of soap. It will blend into the color of the soap, and appears natural. Get someone from the audience to wash his or her hands with water and this soap. The hands become streaked with color. It is so unexpected, it's startling! To get the dye off hands use soap and water and scrub. This is a joke stunt.

6. Dry Water

> You will need:
> a coin
> a bowl
> water
> *lycopodium powder*

This is another trick in which a common article behaves in an unexpected way. In this instance, you introduce "dry water." To prove the point, toss a coin in a bowl of water, and announce that you will remove the coin without wetting your hand. You can do it, as you have secretly rubbed lycopodium powder on your hand. The powder is invisible on your skin, and your hand appears ordinary. Yet, when you dip this hand in the water it will remain perfectly dry. You can reach right into the bowl of water and fish out the coin without wetting your hand. Spectators can feel and see your hand is dry if you show it to them before plunging it into the water. Let them examine the hand before and after.

7. A Flash of Fire

You will need:
lycopodium powder
a candle
matches

Lycopodium powder has another use in magic besides keeping your hands dry. If you place a lighted candle on your table, and from a safe distance, throw a pinch of lycopodium at the flame, it will flash up as if coming from the candle. The flash of fire is spectacular!

8. The Magic Wound

You will need:
iron chloride
a rubber dagger
sodium sulphocyanate

Apply a solution of iron chloride to the back of your hand. Then take a rubber dagger (you can obtain one from a toy store) and dip its blade in a solution of sodium sulphocyanate. You can tell the audience that you are sterilizing the dagger before you perform this "dangerous" experiment. Then draw the edge of the rubber dagger in a cutting gesture over the back of your hand. When it comes in contact with the iron chloride a bright red streak will result. It looks as though you have cut yourself deeply. You can get some screams out of the spectators with this one before the audience realizes that you aren't cut. These chemicals can be purchased from a chemical supply house already prepared as solutions.

9. The Ghostly Faces

You will need:
alcohol
a dish
sodium chloride
matches

Have a group of your spectators gather about your table as you pour some alcohol dissolved with sodium chloride (table salt) into a dish. You can use denatured alcohol (burning alcohol) for this experiment. Just put a teaspoonful of salt into the alcohol in the dish (a semi-saturated solution will work fine.)

While everyone watches, light the alcohol. It will burn with a peculiar-colored flame. Then turn out the lights in the room, and ask the people to look at each other. Each person's face will appear to be of an eerie color—looking ghost-like.

10. The Red Hand

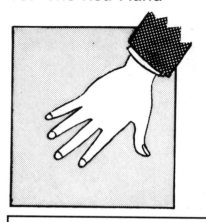

You will need:
  white paper
  sodium sulphocyanate
  a basin
  water
  a water-color brush

Lay a sheet of white paper on the table, dip your hand in water and place it palm down on the paper. When you lift up your hand, a full red palm print of your hand will appear on the paper, as illustrated above.

To accomplish this mysterious effect, make a fairly concentrated solution by dissolving sodium sulphocyanate in water. Use a water-color brush to coat the surface of the paper with this solution. An adult should do this—as sodium sulphocyanate is a chemical to be cautious about, and it should be kept out of the reach of children. When the paper painted with the solution dries, it will look like ordinary paper. When you dip your hand in water and place it against the treated paper, a chemical reaction occurs: When you lift your hand, a big red hand imprint will be on the paper.

11. The Black Hand

You will need:
tannic acid
iron sulphate
a basin
water

This is another trick that, like the above, has a creepy effect. Ask your spectators if they recall the stories of the legendary organization of the Black Hand. Older people will, as it was notorious. Ask someone to explain, or explain yourself that it was a much-feared secret society involved in criminal activities in Sicily and in the United States. Say, "Believe it or not, the Black Hand is still in existence." To prove the point, you dip your hand into a basin of water, and withdraw it as a literal black hand!

The secret of the trick is to rub some powdered tannic acid over your hand (it is harmless). When dusted on the skin, it will not show. Put a tablespoon of iron sulphate into the basin of water. Do this before the show. When you dip your hand in this, it will come out a very black hand. Guaranteed.

12. The Magic Sand of Arabia

You will need:
  fine sand
  wax
  a small box
  a large bowl
  water

While not strictly a chemical magic trick, this feat using treated sand is closely enough associated with this form of science magic to fit nicely into a program of this sort. Heat some fine sand on a stove and add some melted paraffin wax. Stir thoroughly, using enough of the melted paraffin wax so as to coat the grains of sand. By feeling the sand, you can really tell when it is well-coated. Let the mass cool and crumble it into fine particles. It will appear as ordinary sand, yet when you place it in water, it will remain dry.

You can use this treated sand to perform an amazing feat of magic, as the wax-prepared sand will stick together in a lump when you squeeze it. To prepare for the trick, squeeze some of the sand together in your hand to about the size of a Ping-Pong ball. Bury this ball of waxed sand in a box of ordinary sand on your table. Tell the spectators you are going to show them a remarkable feat of magic that comes directly from the desert of Arabia. Then exhibit the box of sand.

Reach into the sandbox and bring out several pinches of the unprepared sand, which you then give to different spectators so they can examine it and see that it is perfectly ordinary sand. Then fill the bowl with water and pour the box of sand into it; the ball of prepared sand goes right along with the regular sand, and is not noticed.

At this point, announce that you will place your hand into the water and will bring a handful of *dry sand*! Let those near you examine your hands, then plunge your right hand directly into the water and stir up the mass of soggy sand. As you do this, you can easily feel the ball of waxed sand; grasp it in your closed fist and bring it forth from the bowl of water. Knead the sand between your fingers as you let it trickle down from your hand onto a paper on the table—dry sand.

The Magic Soda Fountain Act

You will need:
  a glass pitcher
  granulated sugar
  a tray
  glasses
  food coloring
  water

Children love this soda pop trick—it is a tasty surprise.

The Effect:

Display a pitcher full of what seems to be clear water. A tray of empty glasses is brought in, and, one by one, you fill them—from the same pitcher—with different soft drinks, such as lemon, lime, and strawberry. The sodas are passed out to drink, much to the delight of the youngsters.

Here's How You Do It:

The secret is subtle and simple. The pitcher of clear water is really a solution made by dissolving granulated sugar in clear soda water (obtainable at any food market). The glasses on the tray have a few drops of food coloring and flavoring (purchased from any grocery store) in the bottom of each glass.

To perform the trick, all you have to do is pour the "water" into each glass. It combines with the food coloring and the glass becomes filled with whatever soda drink you wish to call it (depending upon the color used). Use some showmanship to build this up, calling out the various flavors you claim to be producing such as: "Here we have lemon; here we have strawberry; here we have root beer," and so on. If you prefer, you may simply use plain water sweetened with sugar for this trick. If you do, refer to the drinks poured as "soft drinks."

You can pass these drinks out to the children to taste, and, thanks to the color and sugary content, they will be pleased.

Chemical Magic Entertainment

You can present your exhibition of Magic with Chemistry either as individual feats or combined into an act. An act of this nature can provide a fine segment for any magic show. For this purpose, select about half a dozen chemical tricks of good variety. Have each trick arranged on its own tray offstage or behind a screen, and have an assistant bring them out and take them away as they are used during the act. Keep things moving at an even pace. You'll find these effects are much enjoyed by audiences. You can open your act in chemical magic with a patter such as this:

"Ladies and gentlemen, I am going to show you a different kind of magic—the magic of chemistry, which produces some very mystifying illusions. This is the real magic from nature's pharmacy that I believe you will find quite puzzling. In fact, I must confess that even as the performer, I too find these mysteries puzzling. And so, I invite you to witness a panorama of mystery featuring effects with liquids."

You can now go into whatever routine of chemical tricks you have arranged. As a conclusion to the act, the following effect is a good one as it is both colorful and mystifying.

The Rainbow Waters

You will need:
  6 aniline dyes
  glycerin
  water
  6 glasses
  a tray

The Effect:

A tray with a decanter full of water and six glasses is placed upon the table. You tell the audience that you have developed a most unusual kind of water that has the unique property of assuming different colors on command. You demonstrate this as follows:

Pick up one of the glasses and fill it from the decanter. The water remains unchanged. You take a sip of it to prove it is just ordinary water. Pour the water from the glass back into the decanter, and announce that this time the water will become crimson. You pour the water into your glass again and it IS crimson. Pick up a second glass and announce that this time the water will be yellow. Fill the glass from the decanter and it comes out yellow. Pick up another glass and call for blue this time. The water comes out blue. When you fill a fourth glass in the same manner, the color is violet.

Now announce that you want clear water, and pour out a glass of plain water. By this time, the spectators will naturally begin to get curious about what kind of water you have in the decanter, so you can hand out the glass of water just poured and let someone taste it. Take back the glass just drunk from, command green water, and pour out green. Pick up the last glass and pour out orange. You have produced six glasses of rainbow water from the one decanter.

Here's How You Do It:

The secret of the trick lies in careful preparation, partly of the glasses and partly of the decanter. To arrange for the trick, purchase some aniline dyes in powder form (Diamond or Rit dyes, obtainable at most stores, work fine for this) of six different colors, for example: crimson, yellow, blue, violet, green,

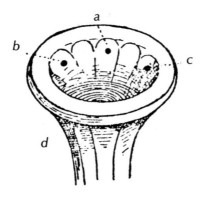

and orange. Mix some glycerin and water in equal proportions, and moisten each dye powder separately with the mixture. Rub it in to the consistency of paste.

The glasses (distinguish by numbering them 1, 2, 3, 4, 5, and 6) are arranged in a row on the tray. Glasses 2, 3, and 4 are prepared by placing a little dab of the required colors at the inside bottom of each. Glasses 1, 5, and 6 are left unprepared. The lip of the decanter is prepared with the three remaining colors, a dab of each color dye-paste being placed at the points marked *a*, *b*, *c*, and *d* as shown in the drawing on page 76. The fourth point, *d*, is left vacant.

The working of the trick will now be obvious. In filling glass 1, you pour over the *d* point on the decanter into the untreated glass. Thus, it comes out clear water and you can sip it. <u>Remember to sip only the clear water, not any water that has come into contact with aniline dye.</u> Pour the water from this glass back into the decanter. The next time you pour water from the decanter, turn it in your hand to a position from which the water will pour out over the dab of the required color, and the glass fills with water of that color.

Glass 2 is now filled from the decanter, pouring the clear water over point *d*. When it hits the dab of color in the bottom of glass the water takes on that color. Glasses 3 and 4 are handled in the same way. Glass No. 5 (the untreated glass) can be given to a spectator to take a drink. Only water that does not touch the dye should be sipped. The next time you fill this glass, allow the water to come out over another dab of color on the mouth of the decanter and it appears to be colored. Then pick up the last glass and show it. Being untreated, it may even be examined by someone in the audience. Now pour the water from the decanter into it, going over the last dab of color, and the glass fills with another color of water. You have magically produced a row of six glasses of rainbow waters from a container of clear water. After the show, be sure to thoroughly rinse out the pitcher and glasses with clear water.

4

MAGIC WITH DRY ICE

Frozen carbon dioxide (CO_2) is known commercially as dry ice. It is easily obtainable—it is used for packing ice cream and any creamery can supply you with blocks of it. It is far colder than regular ice and there are numerous magic effects you can do with this fascinating substance.

To ensure a safe performance while using dry ice, it is important to equip yourself with the proper containers and apparatus. Dry ice carelessly handled can cause a painful burn because it is so cold. Simply use caution, and you'll find dry ice is easy to work with. Use a pair of canvas gloves, and carry the dry ice in a wooden box with a hinged cover and a handle. Insulate the inside of the box with Celotex® (ctyropim). Also build a second, smaller wooden box to fit inside the outer insulated box in which you place the dry ice to be used in each performance. Also, provide yourself with a wooden mallet, a chisel, and a canvas bag in which to pound dry-ice cakes to produce a granular or powdered consistency. Handle the granular dry ice with a large, plastic spoon or tongs. Arrange your apparatus neatly, and you'll develop a professional presentation. Here are some pertinent facts about the production of this chemical, some of which you may wish to include in your patter when you present these demonstrations:

CO_2 is a chemical produced by the combining of the two elements, carbon and oxygen. Chemically, it is produced by an oxidation process whereby the dry ice releases oxygen into the air. The process is such as that of respiration, fermentation, the burning of fuel, and the decaying of vegetable matter. These processes are natural and automatic. In the laboratory, it is achieved by the action of hydrochloric acid on marble chips. The gases produced are absorbed in an alkaline solution, usually sodium hydroxide, which is converted to the carbonate. Carbonate, in this instance, means the chemical sodium hydroxide is converted into CO_2 (Carbon Dioxide) in the process of its manufacture. When saturated, the solution is boiled to drive off the CO_2 gas, which is properly washed, pumped off, and compressed under tons of pressure, finally to be stored as liquid CO_2 in steel cylinders. Further processing produces a fine, solid, snowlike material having a temperature of $-109.3°F$ or $-78.5°C$.

This material is subjected to hydraulic pressure in order to compress it into solid cakes. Dry ice is so cold that it is constantly evaporating. However, the quantity of gas involved is so great and the blocks are so firmly compressed that considerable time is required for it to melt. Actually, solid CO_2 does not melt in the ordinary sense. It evaporates or passes directly from a solid state to a gaseous state, without going through the intermediary liquid stage passed through by water ice (H_2O). CO_2 cannot exist as a liquid at atmospheric pressure. Only under the pressure of one thousand pounds or more may this gas be kept in liquid form.

Solid CO_2 is used as a refrigerant in railroad cars, ships, and stores. It is used for keeping ice cream and other perishable dairy products from spoiling. The fact that it evaporates leaving no residue or moisture is the reason for its nickname of "dry ice."

Try the following patter on your audience. The scientific information adds to the interest:

"I am going to show you some experiments in the magic of science using a fascinating material known as dry ice. Dry ice is among man's wonderful chemical servants. Possibly you are curious as to just what dry ice is, and just how it differs from ordinary ice. Firstly, dry ice is a solidified gas, whereas ordinary ice is just frozen water. The gas used is called carbon dioxide, one of the commonest known to man. Million of tons of it are produced every day by the burning of coal, wood and oil—not to mention breathing. Second, dry ice is 142° colder than water ice. Third, dry ice is much cleaner to handle than water ice because it does not melt but changes directly into a gas when warmed.

"As you can well appreciate, dry ice is very useful for refrigeration, which is its major commercial use today. The invention of dry ice might have died had it not been for the great popularity of ice cream. Dry ice proved a fine way to keep ice cream frozen. Carbon dioxide gas in the air stops the growth of molds on beef, pork, and cured bacon, and similarly prevents the decay of many kinds of fish. Dry ice, thus, will be seen to have great value in performing a double service: keeping perishables cold and in giving off the gas required to prevent mold.

"Hollywood has come to use dry ice to produce spooky fog effects in horror movies. Any density of fog—from a light mist to a pea souper—can be effectively produced by passing steam over lumps of dry ice.

"Another novel use for dry ice is in the trimming of rubber goods. Such articles frequently bend and "give" under even the sharpest knife; however, when they are cooled with dry ice, they become hard and can be cut easily without leaving a rough edge.

"New and practical uses are being found for dry ice all the time. I will now show you some interesting experiments that can be performed with this amazing substance. Get ready for magic with dry ice."

You are now ready to go into your performance. Select the tricks you like best and design your show. You will find that magic with dry ice provides interesting and informative entertainment for all ages.

Making a Dry-Ice Fountain

You will need:
a gallon jug
a cork
glass tubing
rubber tubing
water
dry ice

Find a gallon jug or bottle and fit a cork with a hole in it into the opening. Place a short piece of glass tubing in the hole in the cork, and attach a length of rubber tubing to it. Fill the bottle with water and drop in several lumps of dry ice. The "fountain" is now ready to operate. The dry ice quickly vaporizes in the water and forms gas, which forces the water through the rubber tube. This "fountain" is a striking effect. You can leave the top of the pitcher open for this trick, because the gas is heavier than air and will stay in the pitcher.

WARNING: REMEMBER THAT DRY ICE CAN CAUSE A PAINFUL BURN IF MISHANDLED. BE CAREFUL AT ALL TIMES—INCLUDING DURING THE PERFORMANCE—AND STORE OUT OF REACH OF CHILDREN.

How to Weigh "Nothing"

You will need:
an empty glass
a two-pocket scale
dry ice

Place an empty glass on one side of a two-pocket scale and balance it. Now drop a few lumps of dry ice into an empty pitcher and pour CO_2 gas from the pitcher (dry ice will not harm glass in any way) into the empty glass on the scale. CO^2 gas, although invisible, will pour just like water. The gas, being heavier than air, will displace the air in the glass, causing that side of the scale to go down; but, being absolutely invisible, it appears that you have actually weighed "nothing."

Putting Candles Out by Magic

You will need:
candles
matches
dry ice
a glass pitcher

Light some candles on your table. Drop a lump of dry ice in a pitcher and let it vaporize. Although nothing is seen, as you pour the gas from the pitcher over each candle, each candle in turn is snuffed out mysteriously.

Making Water Boil
from Freezing Cold

You will need:
dry ice
a glass bowl
water

Drop a lump of dry ice in a glass bowl filled with water and watch the water appear to boil. The effect appears exactly as if the water were being boiled on a hot stove. The stunt is novel as it is the reverse of what the audience expects.

A Few "Quickie" Tricks
with Dry Ice

You will need:
dry ice
a bowl
water
water ice
soap bubbles

Place some small lumps of dry ice on the surface of a bowl of water. The lumps will float around like balls of cotton and slowly disappear.

When small lumps of dry ice are placed on a block of ordinary ice, they melt more slowly, and actually melt holes in the water ice.

Place a lump of dry ice on a polished surface and lightly touch it with a small stick. It will seem to glide over the surface as if it were supported on invisible wheels.

Place a few lumps of dry ice in an empty bowl and allow it to evaporate for a time. Now, blow some soap bubbles and carefully drop them in the bowl. The bubbles will float about within the bowl, buoyed up by the dense, invisible surface of carbon dioxide gas, like so many corks in water.

Half-fill a glass bowl with acetone, and drop lumps of dry

You will need:
  a glass bowl
  acetone
  a flower
  a hammer
  rubber bands
  a steel spring
  stiff paper
  mercury
  a pencil
  a nail

---

ice in it. At first the liquid will boil violently, but as more dry ice is dropped into it the liquid mixture becomes quiet and resembles a slush of snow and water. This "slush" is exceedingly cold, having produced a temperature of − 166F° or 198 degrees below the freezing point of water.

You can use this acetone/dry ice solution for many interesting effects. For example, dip a flower in it. The flower immediately becomes frozen solid and may be cracked with a hammer. Rubber bands become brittle, and a steel spring can be frozen to immobility.

An interesting experiment is to make a "mercury hammer" using this acetone/dry ice solution. To do this, make a mold of stiff paper in the shape of a small trough about one-half inch wide and deep, and two inches long. Fill the paper trough with mercury, immerse it in the freezing solution, and hold a wood pencil in the center of the mercury. Do not allow the mercury to touch your skin; it can be very dangerous. It will shortly freeze solid, and you can use the mercury as a hammer. Drive a nail into a board with it, if you wish.

More Novel Tricks with Dry Ice

You will need:
dry ice
a large metal spoon
a half-dollar
a rubber balloon
a board
a bottle
a cork

You can use your own ingenuity and develop good entertainment with this chemical. For instance, if you press a large metal spoon down hard on the surface of a piece of dry ice, it will "squeal" like a pig. Or, lay a half-dollar on a lump of dry ice, and listen to the strange noises it makes. If you use a chisel to cut a slit in the lump of dry ice of such size that the half-dollar will slip into it just halfway, when you place the half-dollar in this slit it will vibrate in a most mysterious way. Place a few small pieces of dry ice inside an uninflated rubber balloon and tie the neck. The balloon will begin to inflate as though blowing itself up. Put a few pieces of dry ice in an empty bottle and cork it. In a few moments, the cork will be blown out of the bottle with a sharp report.

Jars of Mystery Vapor

You will need:
2 wide-mouthed jars
water
dry ice

Dry ice can provide a weird setting for your magic performance, or for a play on stage. Place a wide-mouthed jar on each side of the stage. Half-fill these with water and drop in each a two-inch square piece of dry ice. A fog-like vapor will issue from the jars throughout the stage performance.

Rubber-Band Snakes

You will need:
  a heavy rubber band
  dry ice
  scissors
  a tennis ball
  a frankfurter
  a hammer

Wrap a heavy rubber band tightly around a piece of dry ice, being sure to avoid touching it. Then cut the band with a pair of scissors and let it drop to the surface of the table. It will twist and squirm very much like a snake.

Another stunt combining rubber bands with dry ice involves cutting a large band and showing it as one length of rubber. Hold onto each end of the band and stretch it as far as you can. Then place it into a jar of dry ice and leave it for ten minutes. When you bring it out, take hold of the band by both ends and stretch it out as you did before. The rubber band will now break apart in several places; all elasticity is gone and it breaks like a piece of thin glass. Or try this stunt with a tennis ball. Place the ball in a jar of dry ice so it is completely immersed in the material. After a sufficient time (about 15 minutes), the tennis ball will freeze into a hard, brittle mass. When thrown to the floor with force it will shatter into a thousand pieces. For some extra fun, surround a frankfurter with the dry ice in the jar, and when you take it out smash it into pieces with a hammer.

Soft Drinks for Everyone

You will need:
  flavored drink powders
  water
  glasses
  dry ice

Kids love this stunt. At a grocery store, get some of the flavored drink powders that are so popular. Mix these in water and pass them out to the youngsters to drink. They'll taste good but rather flat. Now, drop a lump of dry ice into each glass. It is fun watching the drinks fizz, and when the children taste their drink after you do this, it proves to be a zippy carbonated soft drink—ice cold!

The Ghost Rings

You will need:
  a tube-shaped box
  balloon rubber
  dry ice
  a candle
  matches
  music

The Effect:

In a darkened room, the spectators feel clammy, ghostly fingers upon their bodies. This is a great spooky party stunt.

Here's How You Do It:

Get a tube-shaped box (the kind that oatmeal or cornmeal comes in). Cut a hole in one end about an inch and a half in diameter. Get some balloons, cut them into flat pieces of rub-

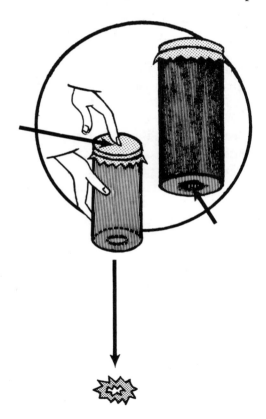

ber, and stretch a sheet of balloon rubber over the other end of the box. Wrap some string around it and tie off to hold the sheet rubber in position over the mouth of the tube. The drawing shows how this is made. This is your "ghost ring gun."

Now, place some broken pieces of dry ice into the "ghost ring gun," and allow it to evaporate for a few minutes until the tube becomes filled with carbon dioxide gas. Hold the "gun" horizontally and point the hole end at anything eight or ten feet away; when you tap on the stretched rubber, an invisible "ring" of cold gas will be discharged and will fly through the air. Hold the "gun" level—rifle-fashion—and aim it at any object you please. The "ring" of cold will hit the object. These are the "ghost rings." If you shoot a "ghost ring" at a burning candle, the candle will go out! If you point the "gun" at any person and discharge it, that person will feel a cold, clammy touch wherever the "ghost ring" touches flesh on arms, face, or neck. The device can be held upside down over someone's head, and a slight tap on the rubber diaphragm will shoot a "ring" downward that will envelope his shoulders, and he will feel a cold wave go through him.

In presenting the effect, tell the spectators that the house is haunted and suggest a séance to see if they can sense the presence of ghosts. You'll need a confederate to operate the gun while you conduct the "séance." All join hands and form a circle about the room. A candle is lit upon the table and the lights are turned out.

Suddenly the candle goes out, and the room is plunged into darkness. At this point, have some spooky music start on your tape recorder or record player. Then the excitement begins as the spectators feel ghostly fingers plucking at them. When the room lights are again turned on, there is nothing in evidence to account for the ghostly hauntings.

Your "ghost rings" produce all the weird happenings, as your secret confederate has the "gun" ready for action. From a hidden position he takes aim, and out goes the candle. Then, in the ensuing darkness, he shoots "ghost rings" at the spectators. The effect is eerie as the chilly rings hit them in the dark.

5

MAGIC WITH PHYSICS

Physics is the branch of knowledge treating of the material world, the science that deals with those phenomena of inanimate matter involving no changes in chemical composition; that is, the science of matter and motion.

This chapter provides many interesting tricks and novelties useful both for entertaining and for learning the principles of the physical sciences. All produce unexpected and enjoyable effects and instruct onlookers in the laws of physics in relation to the fields of acoustics, centrifugal force, gravity, mechanics, and optics. Magic with physics is science at play—or played with.

Acoustics

Acoustics is that phase of the science of physics related to sound and hearing. Sound is heard when any shock or impulse is given to the air, or to any body in contact directly or indirectly with the ear. Sounds travel as waves in much the same manner as the waves seen in water, with a velocity of 1,142 feet per second. Sounds in liquids and in solids travel more rapidly than in air. Two stones rubbed together may be heard in water at a distance of over half a mile; solid bodies convey sounds to great distances, and pipes carry the voice exceptionally well. Here are some interesting experiments with sound:

Visible Sound Vibrations

You will need:
  a glass
  water
  a violin bow
  cardboard
  sand
  a rod
  a 6-inch long dowel,
     three-quarters of an inch
     in diameter

You can make the invisible sound waves become visible by using a glass of water about two-thirds full. Draw a violin bow against its edge, and the surface of the water will show interesting fan-like designs.

Make two six-inch discs of cardboard and connect them with a 3-inch length of ½-inch wood dowel, as shown in the drawing. Sprinkle some sand on the surface of each disc;

then draw a violin bow against the edge of the top disc, which will vibrate and cause the sand to form designs. The same design caused by the sound vibration will be transferred through the rod, and will appear appear simultaneously in the sand covering the surface of the lower disc.

The Musical Bottle

You will need:
   a tuning fork
   a bottle
   water

A tuning fork is very useful in experimenting with sound. Get a glass bottle and pour water into it. By tapping the side of the bottle with a metal spoon to check sound, fill it with water until the sound of the bottle being tapped corresponds to the sound of the tuning fork. Once you have so "tuned" the bottle of water, when you bring the tuning fork near to the mouth of the bottle, it will give out a musical note as the column of air within it vibrates in unison with the fork.

The Musical Glass

You will need:
   a glass goblet
   water
   paper

Use a thin glass goblet for this experiment. Fill the glass almost to its top with water, and lay across its mouth a paper cross, as shown in the drawing. Turn down the ends of the cross at right angles in order to prevent it from slipping.

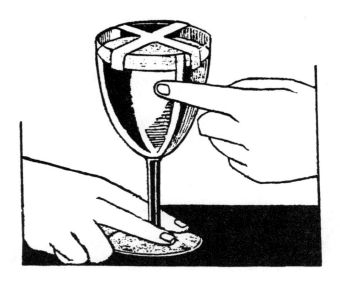

You can now cause the glass to vibrate by rubbing your wet finger on its exterior surface, and it will emit a sound. But, more than this, you will note that if your finger rubs the glass under one of the branches of the paper cross, the latter will remain stationary. However, if you rub a part of the glass in between the branches of the cross, the cross will begin to turn slowly, as if obeying some magical influence. It will not stop turning until the end of one of the branches arrives over the part of the glass rubbed by the finger. By this means, by moving your finger around the glass, you can make the cross turn mysteriously as you please.

This experiment demonstrates the existence of the "points of least motion," which are called in acoustics "nodes" and "ventral segments of vibration." The nodes, where the branches of the cross stop, are the points in which the top edge of the glass are at rest; the ventral segments, situated between the nodes, are the points in which the vibration on the glass's border is the most sensitive, and on which the branches of the cross will not remain at rest.

How to Tune a Guitar Without Using Your Ears

> You will need:
>   a guitar
>   paper

Guitars are very popular musical instruments. Here is an unusual way to tune the instrument by *sight* rather than by sound. Place a saddle of paper (like a capital A; see the illustration) on the second string. Then sound the first string of the guitar loudly and the sound vibration will be transferred to the second string and will make the saddle of paper shake or fall off. Then commence to tune the string at the handle in the usual fashion; when the string is in harmony, the paper will remain still upon the string and will not shake at all. You can proceed, thus, from string to string; and, by a visual observation of the saddle of paper, tune the guitar perfectly.

The Sound Magnifier

```
You will need:
  a rubber balloon
```

Just as a magnifying glass will magnify light, so a balloon will magnify sound. To demonstrate this, take a rubber balloon, blow it up, and tie it off. Have a person hold it to his ear while you whisper toward it from a distance. The magnification of sound is striking and surprising.

The Aeolian Harp

```
You will need:
  thin wood, 22 inches × 48 inches or enough for
  dimensions given
  7–12 catgut instrument strings
  hand tools
  wood glue
  screws
```

Ever hear of an aeolian harp? It is a very ancient instrument that is played by the wind. You can make one.

Construct a long narrow box of thin veneer wood, like a cigar box, including the top, which is fastened in position to form a completely closed resonating chamber. Make this box about six inches deep, ten inches wide, and twenty four inches long. In the center of the top of the box, drill small holes in the form of a circle about two inches in diameter. Passing over this circle of small holes drilled in the top of the resonating chamber run seven to a dozen strings of very fine catgut (violin strings, obtainable at any music store) stretched over bridges (½ inch by four inches) at each end like the bridge of a violin, and screw these strings up tightly as is customary with stringed instruments.

The twelve strings must all be tuned, as you would tune any stringed instrument, to the same note. D, which is just above middle "C" on the piano keyboard, is a good note to use. The upper three strings of the Aeolian Harp can be tuned to D above middle D. The six central strings can be tuned to middle D, and the three lower strings can be tuned to D below middle D.

When ready for the instrument to play, place it in a window that is partly open with the sash raised just enough to give the air admission. When the wind blows upon these strings with different degrees of force, it will produce different sounds. Sometimes the blast brings out all the tones in full concert, and sometimes it sinks the tones to the softest murmurs. It is like music from fairyland, truly a magical instrument.

The Magical Humming Glass

> You will need:
>   a table fork
>   a glass

The Effect:

For this mysterious trick employing sound, you can mystifyingly transfer sound from the tines of a table fork you have plucked (like a tuning fork) to a glass upon the table. The humming of the fork seems to become amplified in the glass.

Here's How You Do It:

Seat yourself at a table with a wooden surface. Place a glass upon the table, and hold a table fork in your left hand. Now, pluck the tines of the fork with your right fingers (see drawing *a* above), and it will hum. Seemingly carry the sound in your fingers and place it in the mouth of the glass, and the hum appears to occur there. Actually it is audio misdirection, for as your fingers apparently place the sound in the glass, secretly touch the handle of the fork, which is still in your left hand, to the wooden surface of the table (see drawing *b* above). This causes the table to act as a sounding board amplifying the sound, which now seems to be coming from the glass. The illusion is remarkably effective.

The Matchbox Monte Trick

You will need:
  3 empty matchboxes
  adhesive tape
  1 full matchbox
  a rubber band

The Effect:

Three matchboxes (each sealed closed with adhesive tape) are used for the trick. Two seem empty and one full of matches. You shake each matchbox in turn, and the full one is clearly heard to rattle. You place them on the table—one by one— and slowly move them about to mix them. The spectators are asked to watch very closely, and see if they can guess where the full box is. It looks easy, but they miss everytime!

Here's How You Do It:

A subtle misdirection in sound is responsible for the success of the trick. Secretly you have a box containing some matches up your right sleeve (beneath your coat sleeve) fastened to your forearm with a rubber band, as shown in the drawing below.

The three visible matchboxes that you show to the spectators are really all empty. Since they are sealed, no one can look inside, so to show whether they are full or empty you have to shake them. Thus, when you want a box to sound empty, shake it with your left hand; when you want a box to sound full of matches, shake it with your right hand, and the box up your sleeve rattles so it seems that the sound is coming from the box the spectators see, and it is assumed to be the full box.

The human ear is very susceptible to deception.

Mixing the boxes and letting the spectators guess which one is full of matches is entirely showmanship. Present the effect as a game of chance. Naturally, since all the boxes are empty, the spectators always miss no matter how closely they watch or which matchbox they guess. Only you have the magic know-how to pick up the "full box" and rattle it any time you wish.

Capillary Attraction

Things floating on the surface of liquids seem to act as though the liquids were covered with a very thin elastic membrane, the power of contraction of which varies with the nature of the liquid. In physics, this is termed the "superficial tension of liquids." The following two fun experiments illustrate this capillary attraction effect.

The Hungry Matches

You will need:
    a basin
    water
    wooden matches
    soap
    a lump of sugar

The matches used in this demonstration in "magical physics" and children have much in common. Like children, the matches flee from a piece of soap but are attracted to sugar!

To perform this stunt, place a basin of water upon the table, and float a group of wooden matches on the surface of the water in a circle formation; place them more or less in the form of a star with their heads near each other. In the center of this star thrust into the water a pointed piece of soap. All the matches immediately begin to swim off away from the soap, as shown in drawing *a* below.

Now take a lump of sugar, and say you will coax them back. Dip the sugar into the center of the water, and the matches quickly move back toward the center, grouping around the sugar eagerly, as shown in drawing *b*.

The Racing Water Drops

Learning these few facts about water drops will increase your fun when you create this effect:

The first drops of rain to fall on a dusty path form little balls and rebound on the soil as if they were elastic. A drop of water on a heated plate takes the form of a flattened ball, for it is protected against the action of heat by the cushion of steam interposed between it and the plate. If you allow a drop of water to fall on a sheet of paper, it will spread out into a large circle; then we say that the water *wets* the paper. But, if you oil the paper so the water cannot wet it, the drop of water will roll on this paper like a slightly flattened ball.

Now that you know about the process of capillary attraction, have fun learning by playing this scientific "magic" game.

> You will need:
>   waxed paper
>   several jacketed hardcover
>   books
>   pins
>   a spoon
>   water

From the roll of waxed paper, take a long strip and cut it about ten inches wide. Then, arrange on the table several jacketed hardcover books[3] of decreasing sizes (see the drawing

[3]If you prefer not to make pinholes in book dust jackets, wrap plain sheets of paper or newspaper tightly around books without jackets. You just need a support for the waxed paper.

on this page), and pin the strip of paper to their jacket spines, allowing little valleys of the paper to fold between each book, as illustrated. Once the end of the paper strip passes over the lowest book, it terminates in a saucer.

Play the game by spilling a few drops of water from a spoon at the point where the waxed paper goes over the largest book. These drops will roll down the inclined plane with increasing acceleration and will rise over the back of the second book until they fall, one after the other, into the saucer.

There is something fascinating about watching the drops of water ascending and descending, and racing one after the other on these miniature mountains. People of all ages love to play this game of racing the drops.

Centrifugal Force

Everyone knows about centrifugal force. It is so universal a phenomenon that even earth itself is flattened at the poles and bulges out at the equator because of it. Now you can play with it.

The Whirling Coat Hanger

You will need:
  a coin
  a wire coat hanger

Get a wire coat hanger and a small coin (penny, nickel, or dime), and balance the coin carefully on the middle of the crossbar. Hook the hanger over your forefinger, and start swinging the hanger gently back and forth. Gradually increase the momentum until the hanger spins in a complete circle around and around your finger, as shown in the drawing on page 106.

If you do it skillfully, the coin will remain on the crossbar even when upside down, held there by this law of physics. To end the experiment, gently bring the hanger back to a stop, and remove the coin. You have demonstrated centrifugal force.

Whirling a Glass of Water Upside Down Without Spilling a Drop

You will need:
  a glass goblet
  water

Thanks to centrifugal force, you can do this startling feat. Place an almost full glass of water on the table before you. Your object is to take the glass of water in your hand in such a way as to cause it to make a perfect circle in the air—with your arm stretched out at full length—and then set it back in its place without spilling a single drop of water.

You can do it, but be careful. It all depends upon the way you hold the glass. Instead of picking it up as if you meant to drink it, pick it up with your hand reversed as shown in drawing *a* on page 108, and then throw your arm out boldly without increasing speed or the slightest interruption in moving it in a full circle, as the arrows indicate. The glass will come to the point, after its sudden revolution, in the position shown in drawing *b*, and from this position you can return it safely to the table.

The Hindu Climbing Balls

You will need:
    a 5-foot pole or sturdy dowel
    fishing line
    2 wooden "bridges," each 3
        inches high
    3 wooden or hard rubber balls,
        2 inches to three inches in
        diameter

This trick is of East Indian origin and is performed by the fakirs. The effect is very deceptive. For the feat, a pole about five feet long is used; stretched along it, from end to end, should be two strings supported by two wooden bridges (more or less violin fashion). Drawing *a* below shows the apparatus.

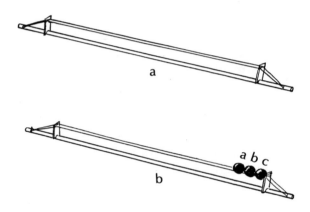

The Effect:

Three wooden balls are placed on the strings against the lower bridge, as shown in drawing *b*. Hold the pole slanting upward and begin to turn in a circle. As you do this, the first ball will climb slowly to the top of the pole, then the second, and finally the third. The balls then, one by one, slowly descend down the pole to their original position against the lower bridge of the pole.

Here's How You Do It:

The trick operates through the use of centrifugal force. First, you must make the equipment. For this, get a pole about five feet long and some heavy fishing line. Cut a groove around the pole near each end and tie the ends of the two strings around each groove. Stretch the strings somewhat loosely along the pole and tie them off. Using a coping saw, now make two small bridges out of wood, each about three inches wide and two and one-half inches high. Cut out a semicircular piece from the bottom of each so that it will fit over the pole. Then, about an inch down from the top of each bridge, cut a slit in each side for each of the cords to pass through. There must be enough of the bridges extending above the cord to keep the balls from rolling off. Place the two bridges under the cords several inches from each end. Study the drawings showing the apparatus and you will understand this construction.

This pole device and three wooden balls about two to three inches in diameter complete the apparatus. Or you can use hard rubber balls such as are obtainable at any variety store.

To perform the trick, hold the pole in your right hand so the pole slants upward. The three balls are placed on the two strings (which form a track for them), and they rest against the bridge on the lower end of the pole.

Support the lower end of the pole against your chest to hold it firmly, and then begin to revolve slowly to the left. Gradually increase the speed of your turn until you start the first ball climbing up the incline slowly, and it finally reaches the top. The drawing on page 111 shows the action.

Continue turning, regulating your speed until you succeed in getting the second ball to join the first at the top, and then the third ball climbs up the string track to join the others.

Now, decrease your speed of revolving, and the three balls will come down slowly, one at a time.

When you perform this trick, be sure you have plenty of room to swing in, and be careful at first not to get dizzy. Practice will enable you to perform the feat without difficulty.

Heat

Heat is produced by causing motion in the molecules of matter, such as is observed when a pan of water is placed on a stove and brought to a boil. Here are a couple of magic-like effects using heat as their modus operandi.

The Dancing Paper Dolls

> You will need:
>   a sheet of stiff paper
>   cardboard
>   a saucepan
>   water
>   a heat source

This effect is for your own enjoyment rather than for a performance, since it requires a stove or hot plate.

Fold a sheet of paper about six inches long several times and draw half of a doll on the top surface of the folded paper,

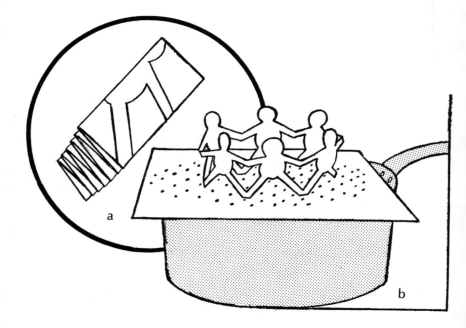

as shown in drawing *a* on page 112. Cut this doll out of the folded paper. The result will give you a string of paper dolls joined together. Paste the two ends together, and you have formed a circle of paper dolls.

Next, punch holes in a heavy sheet of cardboard, large enough to cover a pan filled with water. Place this on a stove and heat it until the water boils. Now place the circle of paper dolls on the perforated cover, and watch them dance, as in drawing *b* on page 112.

The Revolving Snake

You will need:
   thin cardboard
   string
   a candle
   matches

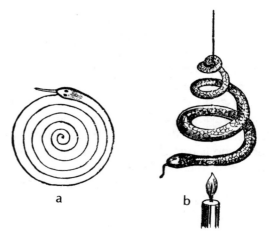

a   b

This fascinating experiment makes an interesting and mysterious toy. The effect operates through heat being applied in a unique way. This effect demonstrates the particular principle in physics known as "obliquity of motion."

Take a sheet of thin cardboard, oak tag or manila, and draw a spiral on it, as shown in drawing *a* on page 113. You can design and decorate it as a coiled snake if you wish. Cut the spiral out and tie a string to the center point of the spiral (the snake's tail). Now, suspend the loosely dangling spiral over a candle flame or gas jet of your stove to heat it, as shown in drawing *b* on page 111.

Amazingly, you will soon find that the action of the heat passing through the spiral will make it spin like a top—revolving around and around with considerable velocity. A very pleasing effect.

Mechanics

The principles of mechanics are founded upon the structure of materials and the laws of motion. The properties of matter are: solidity, divisibility, mobility, elasticity, brittleness, malleability, ductility, and tenacity. The laws of motion are: 1) Every body continues in a state of rest or of uniform rectilineal motion, unless affected by some extraneous force; 2) The change of motion is always proportionate to the impelling force; 3) Action and reaction are always equal and contrary. In the study of mechanics, one learns about the action of gravity, suspension, and other problems of this nature.

Of themselves, presented in such a factual manner, these principles seem difficult to grasp; but offered in the guise of magic stunts, they are not only appreciated but comprehended.

For a simple illustration of the laws of motion, take two marbles and place one of them on the floor. Now shoot the other marble at it, striking it "plumb," as it is called. The struck marble moves forward exactly in the same line of direction. But if it is then struck sideways, it will move in an oblique direction, and its course will be in a line situated between the direction of its former motion and that of the force impressed. This is called the "resolution of forces" in mechanics.

There is certainly nothing remarkable about one marble hitting another, but everyone is familiar with marbles, and this shows that even complicated scientific principles can be illustrated in simple ways. Principles of science taught in this way are long remembered, because a new dimension of meaning is given to everyday occurrences that are henceforth seen as little examples in the visible operation of natural laws.

Gravity

Gravity is the attraction of physical bodies for each other. It is the force that keeps us all on the surface of the earth. Without it, we would all fly off into outer space! "Antigravity" has long been the dream of those who search for the unusual. Levitation is regarded as one of the great mystery themes. It is well to learn something about gravity.

Here's an experiment you can try. Take a stick and balance it upright on your finger; in doing this you are demonstrating a principle of mechanics in relation to the "center of gravity," which is that part about which all of the other parts equally balance each other. In other words, the "center of gravity" is the point at which an object freely acted upon by the earth's gravity is in equilibrium in all positions. The center of gravity principle can be observed in operation in the following little apparatuses, which are quite magical in effect.

The Magic Pipe

You will need:
  a wooden or corncob smoking
    pipe
  a stiff belt

Get an old smoking pipe or a corncob pipe and cut a notch on the side of the bowl, near the stem, as shown in the drawing.

Now, take a stiff belt and place it in the notch you have made in the bowl of the pipe, and let the ends dangle down on each side, as illustrated. The notch, being itself slanted, slants the belt back a bit.

Place the tip of the pipe on the tip of your forefinger, and it will balance there, apparently defying the law of gravity. You can rock it back and forth without it falling. You are showing here the principle of the "center of gravity" in support. If the belt were removed from its slanted position on the pipe stem, the pipe would immediately fall, because the center of gravity would then be in the front of the pipe; but with the belt in position, the center of gravity instantly changes its position—being brought under the pipe—and it again is in equilibrium.

The Prancing Horse

You will need:
  a plastic dimestore horse
  a wire
  a small weight

This toy, a plastic horse in a prancing position, illustrates another way in which the remarkable principle of the "center of gravity" may be used to produce a novel effect. In this instance, the center of gravity is lowered beneath the toy horse by a weight attached to a wire. In this arrangement, the horse will stand upright upon its hind legs and rock back and forth on the edge of a table.

Precarious Balancing

You will need:
a pencil
2 single-bladed pocket knives,
exactly alike
2 corks
a pin
a needle
a bottle
a half-dollar
2 table forks, exactly alike
a yardstick
a bucket
a smaller stick
water

The illustration on page 119 shows three ways in which you can perform some astonishing balancing feats via an altered center of gravity.

Drawing *a* shows how a pencil may be balanced on your fingertip by sticking an identical pair of single-bladed pocket knives in its sides. In drawing *b*, the blades of the same two pocket knives are thrust into the sides of a cork, into which is stuck a pin. This pin is then balanced on the point of a

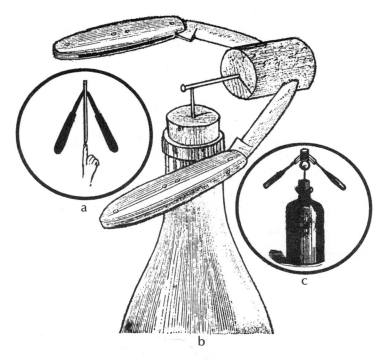

needle supported upright in the cork of a bottle, as shown. The extreme delicacy of balance made possible by the ingenious use of this principle in physics is truly amazing. Drawing *c* shows another experiment, in which a half-dollar is balanced on its edge on top of the point of a needle, using two table forks as the balancing medium. And try this one: Lay a yardstick across a table, letting one third of it project over the edge. You can hang a bucket of water on it without fastening the stick to the table or letting the bucket of water rest on any support.

To make certain this stunt will operate properly and that the bucket will not be able to incline to either side, place a smaller stick inside the bucket at an angle so it pushes up

against the stick lying on the table (as shown in the drawing above). In this arrangement the bucket of water will remain hanging in its unusual position.

The Obedient Egg

You will need:
  a pin
  fine sand
  wax
  several raw eggs
  a bowl

You can use the shifting of gravity to make an effective magic trick in which you show that your "magic egg" will stay in any position you choose to place it. It will stand on the edge of a knife or on the rim of a glass no matter whether you put it sideways or endways. This egg seems literally to defy the law of gravity.

To make the "obedient egg," take a raw egg and blow it out by the usual process of placing a pinhole in each end. Wash it out thoroughly and let it dry. Then fill the blown egg shell a quarter-full of fine sand, and, with a bit of white wax (or some similar material), seal up the pinholes at each end. It looks like an ordinary egg.

Place your prepared egg among others in a bowl, and when it is time to work the trick, pick up this one. You can safely announce that the egg will obey your every command. Place it in any position you wish and it will stay there. The only precaution you need observe is to tap the egg gently each time you place it in a strange position, so the sand it contains will settle inside at the bottom; then it will "obey" you and stay wherever you put it.

Suspension

Here are some puzzles with the mechanics of suspension, posed for you in question form.

You will need:
   16 large wooden matches
   a soda bottle
   a soda straw
   3 table knives
   2 smooth sticks
   2 hardcover books
   cardboard
   glue

Can you lift fifteen matches with one match? The drawing shows you how.

To perform this stunt, use large wood kitchen matches. Lay one match on the table (*c*) and crisscross all of the other matches astride it (*a–a*) with the exception of one (*b*). Place this last match (*b*) above matches (*a–a*) along the angle formed by their interplacement. You can now lift up the one match (*c*)

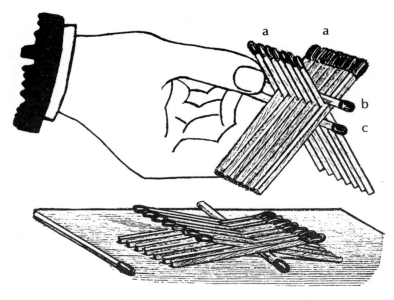

by its end, and the 14 crisscrossed matches (*a–a*) will assume an oblique position, holding match (*b*) within their angle. You can now lift all fifteen matches on the one match!

Can you lift a soda bottle using only a soda straw in direct contact with the bottle? Give up? You can do it if you know the principles of suspension.

Bend the straw upward at a point approximately one third of its length from one end. Put the bent part of the straw into the neck of the bottle. The straw in the bottle will spread out as shown in the drawing, making it possible for you to lift the bottle by the straw alone, and you have solved the puzzle by mechanics.

Can you take three table knives and build a bridge of them that will be self-supporting, and will suspend a weight? Impossible? Not when you know the mechanics of suspension.

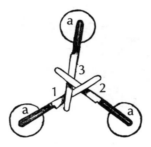

Follow the above drawing to learn how to do this. Place three glasses upside down on the table in the form of a triangle (*a–a–a*). Then arrange three table knives on them, as shown. The blade of knife No. 1 goes over that of knife No. 2, which then goes over No. 3 which rests on No. 1. The bridge of knives so made will be self-supporting because of the "magic" of mechanics.

Can you make an object roll uphill? Using tricky applications of gravity and suspension you apparently can. Construct this apparatus:

Arrange two smooth sticks on the backs of two books of different sizes, so you have an uphill incline; place the sticks so they form an acute angle whose apex falls beyond the smaller book (see drawing). Or, if you wish to make a more formal apparatus, you can construct a permanent set of uphill tracks on a wooden base.

Now make two cones of cardboard and glue their mouths together, which gives you a rolling object in the shape of a double cone. The following drawing shows the apparatus.

Place this double cone near the lower angle of the rods that form the tracks, close to the bottom of the inclined plane. To everyone's astonishment, instead of seeing the double cone roll downhill as expected, they will see it appear to roll uphill.

Optics

Optics is that phase of physics dealing with light and vision. Light follows the same laws as gravity, and its intensity or degree decreases as the square of the distance from its source increases. Thus, at a distance of two yards from a light bulb, we will have four times less light than we would have were we only one yard from it, and so on following the same proportions.

Objects that allow the rays of light to pass through, such as water or glass, are called refracting media. When rays of light enter these, the rays do not proceed in straight lines but are refracted (bent out of their original course). You can illustrate this principle by performing the following trick.

The Invisible Coin Trick

You will need:
   an opaque bowl
   a coin
   water

To do this trick, place a coin in the bottom of any empty, opaque bowl, and have a spectator stand back from it just enough so that he or she cannot see the coin over the edge of the bowl. In other words, the coin is hidden by the edge of the bowl. Now, pour water into the bowl while the person watches from the same position, and as the water rises the coin will become visible and will appear to move from the side to the middle of the bowl.

The Prism Rainbow

```
You will need:
a prism
white paper
a light source
```

Another basic fact of optics is that normal light is composed of a variety of colors in combination. White light can be broken down and its components observed using a prism, which is a triangular piece of solid glass. When a ray of light passes through a prism, it is divided into its three primary colors—red, blue, and yellow—as well as the four secondary colors—violet, indigo, green, and orange. (These colors are called "secondary" because they are made by some combination of the "primary" colors.) The best way to observe this scientific fact is to allow a ray of sunlight to come in through a small slit in a window blind and fall upon your prism. Let the beam of light that passes through the prism then fall upon a sheet of white paper, and a beautiful array of colors will be seen, the seven colors of the rainbow with which we are all familiar.

Further, the colors each cover varying proportions of the color spectrum. If the spectrum were divided into fifty parts, red would cover six parts, orange four parts, yellow seven parts, green eight parts, blue eight parts, indigo six parts, and violet eleven parts. You can prove this with a simple, easy to make device.

The Color Disc

You will need:
  cardboard
  a pencil
  a compass
  paints
  a long pin

Cut out a six-inch disc of cardboard, divide it with pencil lines into fifty equal parts, and paint colors in them in the proportions given. The result will approximate this drawing:

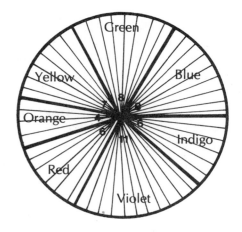

Now, place a pin through the center of the disc so you can spin it, and you will reverse the color division process of the composition of light. As the disc revolves more and more rapidly, the colors will no longer be separate and distinct, but will gradually become less visible, and will ultimately appear as white—the exact color of the light as it originally appeared before being broken down into its colors by prism action.

The eye has some unique characteristics it is well to understand. Try these experiments, which are also interesting to explain to spectators.

The "Thaumatrope"

You will need:
  a flashlight
  cardboard
  a pencil
  string

Ever hear of a "thaumatrope?" The word is derived from two Greek words meaning *wonder* and *to turn*. It is based upon this principle of optics: An impression made upon the retina of the eye lasts for a short interval after the object that produced it has been removed; this impression that the mind receives lasts for about an eighth of a second. You can demonstrate this effect by revolving a flashlight around in the air, and when it gets whirling fast enough you see not one point of light, but a circle of light. This is the optical principle upon which

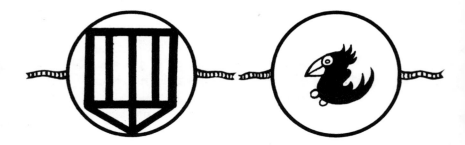

motion pictures are based. You can use it to make an interesting toy.

Following the drawing on page 128, which is enlarged, cut a piece of light-colored cardboard about the size of a nickel, and draw on one side of the disc a bird and on the other side a cage. Fasten two pieces of six-inch long string to the disc, one on each side opposite each other. Grip the ends of the strings, one in each hand. By twirling these strings between the thumb and forefinger of each hand, you can make the disc revolve rapidly, *and you will see the bird within the cage.*

The Blind Spot

You will need:
2 dimes
a blackboard
white chalk

One of the most curious facts about human vision is that there is a "blind spot" in each eye. Any image falling on that particular spot in the retina will be invisible. When we look with the right eye, this point will be about fifteen degrees to the right of the object observed. In other words, it is to the right of the axis of the eye, or the point of most distinct vision. When looking with the left eye, the point will be as far to the left.

You can demonstrate the blind spots in the eyes by placing two dimes on a table. Space these about three inches apart. Then, look at the left-hand dime with your right eye, at a distance of about a foot—keeping the eye straight above the dime and both eyes parallel with the line of the dimes. Keep your left eye closed, and suddenly the right-hand dime will disappear from sight. A similar effect will take place if you close your right eye and look with your left.

Another way to demonstrate "the blind spots" is to draw in the center of a blackboard (use white chalk) a two-inch central spot; about two feet on each side of this and slightly lower down make two other chalk marks, as shown in the drawing below.

Place yourself directly in front of the large center spot and hold the end of your forefinger before your face, so that when the right eye is open it will conceal the little mark on your right, and when the left eye is open the mark on your left. Having obtained this position, look with both eyes at the tip of your finger, and the large central spot will disappear.

The Water Mirror

You will need:
  a glass
  a spoon
  water
  a light source

Nearly fill a glass with water and place a spoon in the glass. Stand with your back to a window through which light is entering, and hold the glass above the level of your eyes, as shown in the above drawing (eye position is at E). (For your understanding, in the drawing, line A–B shows the position of the spoon within the glass and line C–C the surface of the water.) Then look up obliquely following the direction of the dotted line, and you will observe the entire undersurface of the water shining like a mirror. And, as you observe the spoon in the glass in this manner, it will be reflected in the surface of the "water mirror," but with a brilliancy far superior to that seen in any ordinary mirror.

The Boundless Menagerie

You will need:
  heavy cardboard
  flat mirrors fitted to the inside
    walls of the box
  plastic dimestore animals
  glue
  frosted white plastic

Make a box (cardboard will do) about six inches long and wide, and twelve inches high. Cover the inside walls of this box by taping four flat pieces of mirror (which will make them perpendicular to the bottom of the box and to each other). On the inside bottom of the box, glue in various positions little plastic animals that are easily obtainable at toy shops and dimestores. They will provide a wonderful menagerie of all kinds of animals nicely formed in miniature.

Now get a sheet of frosted white plastic and use this to form a lid for the box. Being semitransparent, it will flood the interior of the little chamber housing the tiny animals with a diffused light. Cut a small opening in the center of this covering to the roof of your box so you can peep inside.

When you gaze down into this box, the endless repetitions afforded by the reflections of the mirrors lining the walls of the box will produce repeated images of the animals on into infinity in all directions, producing a "boundless menagerie."

The Magic Mirror

You will need:
  thin wood, about 30 inches ×
    45 inches cut into squares 15
    inches × 15 inches (have
    ovals cut in 4 of the squares
    as described)
  2 mirrors, about 21 inches ×
    15 inches
  glue

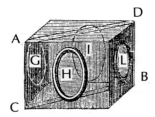

Make a cubical box with sides about fifteen inches long. In each of the four sides of the box, cut an oval opening ten inches high and seven inches wide, as shown in the drawing (C, I, H, L). Inside the box, place two mirrors with their backs against each other. These cross inside of the box on a diagonal line (45° placement), as shown in the above figure (A–D, C–B). Attach the top of the box, so that you have a solid cube with an opening in each side. When you look through any of the openings, you will see the surface of the mirror.

Place this "magic mirror" in the center of a table, and have four persons group around it, so that each can look into the oval opening in front of them into the mirror. To the wonderment of everyone, instead of seeing their own reflection in the mirror, they see the reflection of the person seated next to them.

Because the box is sealed, there is no way for them to examine the arrangement of the mirror, so it is a good mystery. What makes it look like magic is that while rays of light may be turned aside by a mirror, *they always appear to proceed in straight lines*.

The "Seebackascope"

You will need:
   heavy cardboard
   a mirror, 2 inches × 1 inch
   glue

Mirrors produce very interesting optical effects, and have been used in many ways by magicians. A simple device in this regard is called the "seebackascope." With it you can see what is going on behind your head. It is easy to make and it is a lot of fun seeing what goes on behind you.

To make a "seebackascope," construct a little box of cardboard about 1½ inches wide, 1½ inches high, and three inches in length. It is open at one end and on one side, and a small piece of mirror is placed within it at a 45° angle, as shown in the drawing on page 134. When you look into the instrument, it will give a surprising range of vision behind you.

The Periscope

You will need:
  a tube
  2 mirrors, 2½inches × 3½
    inches
  glue

The "seebackascope" is an early form of the periscope later developed for submarines, so sailors can "up scope" and see what's happening on the surface. You can make a periscope any size you wish. Construct a square tube of cardboard. About 3½ inches square and 18 inches in length is a convenient size. An opening is made in the wall of the tube, at opposite ends and on opposite sides, and behind each opening is mounted a mirror (3½ inches square for a 3½ by 3½ by 18 inch periscope) at a 45° angle. The mirrors are mounted in opposing directions to each other, as shown at *a* and *b* in the drawing on page 136. With this instrument you can look around corners, over the edge of a roof, and so on.

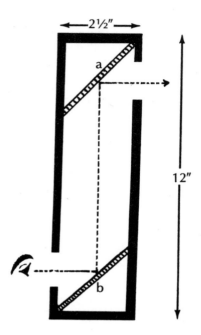

How to See Through a Solid Brick

You will need:
  heavy cardboard
  4 small mirrors, 2½ inches ×
    12 inches
  glue
  a brick
  a support

    The trick requires a hollow, U-shaped box. This U-shaped box can be made either of wood or of heavy cardboard, but it must be solidly constructed to handle the weight of the brick used in the trick). The drawing on page 137 shows how this

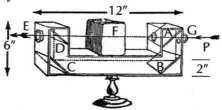

is constructed; one side of the drawing is purposely removed to enable you to see the arrangement of the interior. A, B, C, and D are four small mirrors, approximately 3 by 3 inches in size. These mirrors are mounted periscope-fashion at each end of the device, which should total 12 inches in length and 6 inches in height. The apparatus is mounted on a little pedestal of some kind to make it easier to use. See the drawing. E and G are two openings to look through on each end of the U-shaped box. If you look through opening (G) in the direction of the arrow, you will see whatever is at opening (E) on the other end. However, your vision is actually being carried—via the four mirrors placed as illustrated—around the solid brick which is stood in the center of the device. The illusion is very interesting, as it appears that you can look directly through a brick or any solid object. (Support your apparatus on a pedestal of some kind to make it easier to use.)

The Decapitated Princess Illusion

The use of a 45° angled mirror has produced many "magical" illusions, since it will project a reflection at right angles. A good example of this is the popular "Decapitated Princess Illusion," the effect shown in the illustrations on pages 138 and 139. In this illusion, the head of a girl (let's call her a princess) is shown in a curtained recess. It reposes upon two swords lying across the arms of a throne-like chair. The head is very much alive.

This effect requires elaborate apparatus and is probably rather outside the scope of your performance, unless you happen to

have an extra plush chair, a large mirror cut to fit the chair, and 2 suitably fancy but not-too-sharp swords around about the house. But it's fun to know how it's done anyway. The illusion is accomplished as shown in the next drawing, which presents the modus operandi. The chair is upholstered in red plush, and is placed close to the curtain at the back of the recess. In the back of the chair is an opening placed just below the level of the top of the chair arms. This opening is not seen from the front but concealed by a large mirror placed between the arms of the chair at a 45° angle (see arrow in the drawing below). The ends of the mirror rest in the folds of the fan-shaped upholstering on the inside of the chair arms. The lower edge of the mirror is resting on the bottom (seat) of the chair, and the upper edge is concealed by laying one of the swords along it. As the 45° angled mirror reflects the bottom of the chair (chair seat) the impression is produced on spectators that they are looking through to the back of the chair with an unobstructed view. There is an opening in the rear curtain directly opposite the opening in the chair's back, through which passes a sloping board. One end of the board rests on the rear

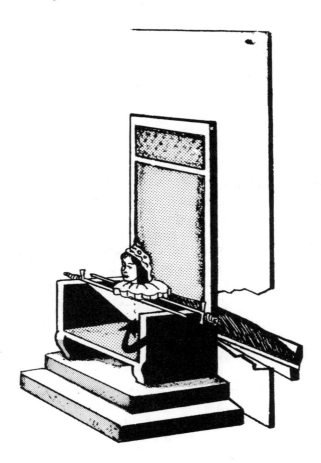

part of the seat of the chair and the other end is supported on convenient legs. This sloping board is used to support the girl's body as she lies behind the mirror, as shown.

The girl who is to be the "Decapitated Princess" takes her position on this board with her chin just above the edge of the mirror: the second sword is placed at the back of her head, and a wide lace collar that she wears is adjusted to drape neatly on the two swords. The effect is startling, as it appears exactly as though a living head were before the spectators, reposing on two swords lying across the arms of the chair. The optical principle is very simple, but the "execution" of this illusion is complicated.

A Black Art Magic Show

The fact that the color black, under certain conditions, will become invisible against a black background is another optical principle that can be used for magic. The principle was originally developed by the famous European magician Buatier De Kolta during the 19th century, and he called it "Black Art." De Kolta made it into a full stage production. The classical show using this principle of optics will first be described below, and then a portable version of the illusion will be given that you can easily build and perform under almost any conditions.

The Classic Show

In the original form of the illusion, the entire stage is hung with black curtains, and even the floor is covered with a black cloth. The regular stage lights are turned off, and a special set of lights lining the lower front and both sides of the stage are employed. The unusual feature about these lights is that they face out toward the audience rather than in toward the stage. These are known as "blinder lights," and have the effect of confusing the eyes of the audience sufficiently so that the stage becomes a chamber of intense *blackness* in which anything black in color is invisible and things white in color stand out in striking contrast, as shown in the drawing on page 141. It is on this special stage—designed as a giant optical illusion— that the mysteries of "Black Art" occur.

In a stage performance of "Black Art," the house lights of the theater dim down gradually and finally go out completely as, simultaneously, the "blinder lights" lining the front and sides of the stage come up to full brightness. The curtain rises, disclosing the stage as a black chamber. In a moment, the magician appears, dressed in a white suit (or a white Oriental costume, as the case may be). The magician waves his hand and a white wand appears floating in the air, which he secures.

A wave of the wand and a white table of the kind in the drawing, appears on stage right; then a second, similar table appears on the left. A large white vase appears on one of the tables, and a second vase appears on the magician's outstretched hand. Both of the vases are shown empty, and in one is placed a few orange seeds. The wand is passed over the vase, which instantly becomes filled with oranges. The oranges are poured into the second vase, they vanish and return again to the first vase, when they disappear as quickly and mysteriously as they appeared, and the vases are again shown empty.

The vases are then placed one on each table. A borrowed watch is placed in one of the vases, from which it disappears and is found in the vase on the other table. A life-size skeleton now appears and dances around the stage, becomes dismembered: the separated parts float about, but they finally rearticulate themselves, and the skeleton then vanishes. Now a stuffed toy rabbit is seen in one of the vases, from which it is taken by the performer, and in his hands it becomes two. The two rabbits are tossed in the air and disappear.

The number of possible tricks performed in the mysterious black chamber are almost unlimited, but an explanation of the ones mentioned in this routine will suffice to show how the optical illusion of "Black Art" is accomplished.

As has been mentioned, while the stage is draped in black, everything that appears is painted white, and the magician is attired in white. There is an assistant on the stage all through the act, but because he or she is dressed in black, with black gloves on hands and head covered by a black hood the assistant is not perceived by the spectators: Black against black becomes completely invisible when viewed through the "blinder lights" shining out at the audience.

The tables are on stage at the start of the performance; by being covered with pieces of black cloth, they are rendered invisible. To make them appear, the invisible assistant has only to pull off the black covering and they seem to be produced magically at the magician's command.

The vases are also on stage, covered with black cloths. The assistant can cause them to appear by removing the black cloth coverings, one on the table and the other on the performer's outstretched hand. The oranges are in a black cloth bag from which the assistant pours them into the vase. To cause the oranges to vanish, the magician, instead of pouring them into the vase, pours them into the open mouth of a black bag held by the assistant just behind the vase. The transposition of the watch from one vase to the other is just as easy. The assistant merely removes it from the vase in which the performer placed it and places it in the second vase. The manipulation of the rabbits is equally simple. The assistant places the first one in the vase by means of a black bag in which it is concealed, then places the second rabbit in the magician's hands from a second small black bag. To cause the rabbits to vanish, the magician merely tosses them up into an open-mouthed black bag held by the assistant, and they appear to vanish in midair.

The dancing skeleton is made of wood sections painted white. It is mounted on a thin board covered with black cloth on its backside. One arm and one leg are jointed so as readily to be moved and put back in place by the assistant when he or she is operating the skeleton.

You now know the modus operandi of "Black Art." Other than careful routing and noise control, such a show does not have many pitfalls, and the effects achieved are among the most magical in the whole field of legerdemain. The illusion is absolutely startling. Yet it is very seldom seen on stage. The reason is obvious: It is very expensive to drape an entire stage in black curtains and carry the necessary lighting equipment; and the stage facilities required for the performance are not too easy to come by these days.

Miniaturized "Black Art"

To meet modern requirements for a relatively inexpensive production of "Black Art" that can be shown in almost any situation, here is a portable form of the illusion using a black-curtained cabinet, six feet by six feet by six feet, for the darkened chamber instead of a full stage. In this smaller form it is much more practical for you to build and present.

You will need:

two assistants

sufficient sturdy pipes and joints to construct a 6 feet × 6 feet × 6 feet portable frame "cabinet"

heavy black cloth with a dull finish to curtain cabinet

6 "blinder" lights with reflectors

an 8 feet × 6 feet bright, thin cloth curtain

a table for props

soil

a large seed

a rose bush or other large plant in flower

a pitcher

water

a tin basin to fit inside a large bowl (black plastic can be substituted) *with handles*

white:

costume for visible assistant

"Oriental" costume for magician

wand

2 taborets (small tables)

flowerpot large enough for flowering plant

cloths

sheet

large bowl

black:
    costume for "invisible" assis-
        tant
    paint for tin basin (if black
        plastic container with han-
        dles cannot be found)
    sheet-sized cloth

The "Black Art" Chamber is made of pipe, a portable frame cabinet that can be quickly assembled. This cabinet is hung with black cloth. Velveteen is the best cloth to use, but any black cloth with a dull finish will suffice. Black canton flannel works very nicely and is relatively inexpensive. Also, the floor upon which the cabinet rests is covered with black cloth. For a cabinet "Black Art Act," such as this, the number of "blinder lights" may be cut down, a bank of three lights on each side of the cabinet being sufficient. These can be mounted on boards readily fixed to the front of the cabinet on each side, facing the audience. Reflectors are mounted behind each bulb, so all of the light goes out toward the spectators and none enters the cabinet. Experiment a little with these lights until the illusion is effective but yet the lights not harsh on the eyes of the viewers. Consider using red-colored lights.

Along the front of the cabinet, on the top pipe, hang a brightly colored curtain of thin cloth. This is arranged on curtain rings so it may be opened or closed as desired. This completes the construction of this minaturized "Black Art" stage, as shown in the drawing on page 146.

The invisible assistant who is to help in the act (any boy or girl can do this) is dressed entirely in black, with a black sack over his or her head in which peepholes have been cut. These peepholes are covered with black mesh so that the assistant's

eyes will not show. Black socks are worn on the assistant's feet and black gloves on his or her hands.

In addition to the invisible assistant, you can use another assistant to help you in the illusion. This second, visible assistant can be dressed in a white Oriental costume and can hand you the various props needed during the show while you are standing inside the cabinet. These props are on a table set outside of the cabinet, over to one side.

First will be described the act as it appears to the spectators, and then the details of its performance will be covered. Have the six-foot-square cabinet set up in the center of the stage or at one end of a room (as the case may be) ready for the performance. The colorful front curtain of the cabinet is closed. This curtain will also function as a nice background for showing previous magical stunts should you have some tricks you wish to demonstrate to your audience before this act is presented. Ready for the "Black Art Act," you are dressed in white: white coat, trousers, shirt, and shoes.

You can present the cabinet as your "Black Mystery Chamber from India" or whatever. Switch on the "blinder lights" (these are mounted in series so that one switch turns them all on simultaneously), pull the front curtain of the cabinet over to one side, and a complete "Black Art" stage is set to go. The show is ready to roll.

Call out some "magic formula" to summon your assistant. Your visible assistant, say a girl dressed in a white Oriental costume, instantly appears. She comes out of the cabinet and takes her bow.

You step within the cabinet, make a wave in the air, and a white wand from nowhere is seen floating in the air, and thence into your hand. A wave of the wand and a taboret (a small, portable table or stand) appears on the right side of the cabinet. Then a flowerpot appears on the taboret, and this is shown to be filled with earth. A seed is planted in it, and the pot is covered with a white cloth, which the girl assistant, from outside of the cabinet, hands you. When the cloth is removed, a rose bush is seen to be growing in the pot. The flowers are cut off and given to the assistant, who passes them out to the spectators.

Next, a large bowl appears in your hand. You wave your free hand, and a second taboret appears on the left side of the cabinet. You place the bowl on the taboret. The girl hands you a pitcher of water from which you pour water into the bowl, completely filling it. The bowl is carefully picked up and the contents tossed out toward the audience—the water is *gone!*

To climax the act, invite the assistant to enter the cabinet while you step outside. You hand her a white cloth or sheet, which she holds up in front of herself. She moves the cloth about inside the cabinet, ghost-like. You recite another "magic formula" and make some dramatic passes with your hands toward the sheet, and the cloth flutters to the floor. The assistant is *gone*!

You pull across the front curtain (closing the cabinet), snap off the lights, and, advancing, call for the assistant.

She comes running down the aisle (or in from a door at the rear of the room) shouting, "Here I am!", concluding a very effective performance of magic.

Here's How You Do It:

As you know the modus operandi used to accomplish "Black Art," you can easily follow how these various effects are produced. It is important to make certain that the invisible assistant never passes in front of you while the two of you are moving about on stage.

The assistant dressed in the white Oriental costume is standing in the center of the black cabinet covered by a black cloth. Your invisible assistant pulls the cloth off her as you call out your formula. She becomes visible.

The appearing and floating wand is simply a white stick that the invisible assistant uncovers and then moves around so that it appears to float in the air, being very careful while holding it (as in handling all white props) that the black gloves do not show against the white wand. You grasp it out of the air. The taborets at the right and left side of the cabinet are already in position, covered with black cloth. To make them appear at the required times, the assistant simply snatches the cloths aside. The flowerpot on the top of the taboret is produced in a similar manner.

To make the bush of roses appear in the pot, proceed as follows: While you hold a white cloth in front of the empty pot, have the invisible assistant exchange, behind the white cloth, this pot for one with flowers growing in it. This is easily accomplished under cover of a piece of black cloth.

The large white bowl that appears in your hand is covered with a black cloth and is placed there by the invisible assistant; the cloth is pulled aside and the "production" is accomplished. To make the water vanish from the bowl, a tin basin is made to exactly fit inside the bowl, as an insert. The basin should have handles on it and be painted dull black, as shown in the drawing. It is this insert that you fill with water. As you pick up the bowl, you can purposely spill a little of the water to prove it is still there. At this instant, the invisible assistant lifts the black basin out of the white bowl, and when you make the motion of throwing at the audience everyone will duck. It is a startling effect!

BLACK FALSE
TIN BOWL

WHITE CHINA
BOWL

Making the visible assistant vanish is easily accomplished by "Black Art." As she holds the white sheet in front of her, the invisible assistant tosses a black cloth over her. When you call out your magic formula, she drops the white cloth, which falls to the floor. It appears that the assistant is gone.

Making the visible assistant reappear from the back of the theater, can only be accomplished when you perform on a stage or in a place with convenient exit door available so she can run around outside the room and enter via another door at the rear of the audience. When the illusion can be done, arrange your cabinet in front of an opening of the rear curtain (if you are performing on a stage) or in front of a doorway (if in a room). As the girl holds up the white sheet before herself, the invisible assistant holds it instead and keeps it bobbing about while the girl slips out through the back of the cabinet and goes around behind the audience, all ready to put in her surprising reappearance. When you make your dramatic gestures, the invisible assistant lets the sheet drop to the floor. The girl seems to vanish on the instant! It makes a wonderful climax for your act when you call her and she comes running down the aisle from the rear of the audience. You take her hand in yours when she arrives back on stage, and you both advance forward for the final bow.

6

TEACHING CHILDREN THE ART OF MAGIC

Your interest in magic will brush off on children, and the time will come when they will take a sincere interest in the art, and will want to perform magic themselves. When the magic bug bites, "conjuringitis" can develop: A severe case will last a lifetime! So when you take children into your confience and show them how the tricks work, you must do more than that; you must teach them a real appreciation for magic; you must teach them the art in magic.

When an interest in magic blossoms, you can teach children how to perform magic *effectively*. Magic is learning, and teaching children magic will provide them not only with good entertainment but with a grasp of some basic science too. Likewise, performing magic cultivates qualities of self-confidence and assurance in budding men and women.

One of the most important lessons that conjuring can teach is that everything in life has a dual character; in every force lies the power for both construction and destruction. The fire that warms your home can destroy your home. The water that quenches your thirst can drown you. Thus, the potential for both good and bad exists in the same things. So it is with performing magic: All depends on whether you use it in an entertaining and skillful manner, or whether you abuse it and perform it poorly. Used properly, the art of magic can bring you success and happiness. Used improperly, it can bring just the reverse. Your audience, if pleased, will applaud you. The same audience, if displeased with your appearance, your manner, your language, or your performance can hurt you by not responding or by rejecting you. It is up to everyone who learns magic, who would like to perform magic, to learn how to please their audience. This lesson goes far beyond just the performance of magic tricks; it applies to everything one does.

People want illusion. People want to escape from drab reality. They rush to theaters, concerts, and special sport events; they sit for hours in front of television sets and watch motion pictures, all to be entertained in an unreal world of fantasy. They want to live in their imaginations and forget for a while life with its routines and its problems. Entertainment allows them to retreat or withdraw from everyday life and see things that are nearer the ideal. In the theatrical world, all is illusion. There, people live through the troubles, joys, and experiences of the characters portrayed. For the duration of the entertainment, if the illusion is successful, each person in the audience forgets his or her own troubles and submerges his or her feelings into the imaginary ones being presented. Paradoxically, this submergence may even help people in the audience achieve more perspective on their own lives.

This is a great factor in the explanation of the power of magic. Magic deals with the unreal. The art is so removed from everyday life that the spell cast by the magician is tremendous. The magician's world appears to be created entirely of make-believe. He or she produces illusion, and that is what people want, *and need*.

Mystery and the supernatural interest people intensely. These will never lose their fascination; because they are the very stuff of which magic is made, the magician will never fail to hold people spellbound. Magic is a great art, and the magician, with the assumed power to produce supposedly supernatural effects, will continue to capture the interest of everyone.

The character of magic, like that of everything else in life, has changed with the ever-changing world. In very ancient times, the magician functioned as priest and sage, and his effects were taken as miracles. Today the art of magic is entertainment that fills the need for fantasy. There was a time in the history of magic when magicians performed with a great deal of cumbersome apparatus: large cabinets and trunks, and long table drapes. The magician, too, was encumbered with voluminous robes. Then came the era where gaudy trappings were discarded and magicians appeared in formal but normal attire. Complicated apparatus was placed aside for less paraphernalia, and skill in sleight of hand was developed.

This was the beginning of the development of impromptu magic, which in recent years has become so popular. The demand for the magician is greater than ever; but the magician does not necessarily perform on a stage, does not require elaborate equipment, does not have special costumes or formal dress. Today, the magician is ready to perform and entertain people under all conditions.

This simplification of paraphernalia and of effects has made the art of magic even more astonishing, by reason of its casual and apparently spontaneous nature. The stress in modern magic is on novelty, surprise, spontaneity. *That is what the modern magician must give.*

The modern magician must attune his or her individuality to the times, and these are times of complete naturalness. Let good taste and modernity be your guide in impressing your audience, whoever and wherever it may be.

The first impression your audience forms of you is very important to your success. You must convey the impression of high standards in your work. Your personal appearance must be neat. If you enter before a group, do so briskly and with confidence. Do not notice people as you enter, wait until you are fully before them. And, if the situation is such that your performance of magic is set for a certain time, do not attract undue attention to yourself in advance. Save that for the psychological moment when you must make the audience notice you and keep their attention focused on you during your performance.

Plan the opening of your magic performance very carefully, for remember, you will sell yourself or harm yourself right at the opening of your show.

Keep your eyes on your audience most of the time. Exert your personality upon them. Talk directly to them. Now and then you may pick out one or more spectators and direct your attention to them. Talk convincingly to these few spectators and impress them, and you will find that this is a means of impressing everyone in the audience. Speak distinctly; speak loudly enough for everyone to hear you and give enough force to your words to send them straight to the mark. Be careful of your language. Use judgment in what you say, and, above all, speak correctly. Put expression into your voice and face. Put life into your performance. Your purpose is to arouse and hold the interest of your audience. Be snappy in your work and get directly to business. Avoid long, slow, drawn-out presentations. Know when to quit.

Study what the public demands; that is, what people want you to give them. The wise magician is like the wise business person, who studies the times and provides people with the kinds of things they are clamoring for. Capitalize on this, and present your magic in attractive packages.

Present your patter and your effects in such a way that they come within the experience of your spectators. Do not do things or say things that are absolutely foreign to their lives. Place your material so it is understood easily by everyone. You will thus capture the interest of your audience and the success of your performances will be assured. Always support magic and other magicians. In every way, be a credit to the art of magic.

When you learn these lessons yourself, both for magic and the performance of magic, you can teach them to children. And they are important lessons that really apply to all of life. The happiest people never stop learning and take an interest in things all during their lives, and, remember, magic is learning.

INDEX OF THE EXPERIMENTS

Add a Number, 32
Aeolian Harp, 97
Age and Loose Change Trick, 32
Amazing Finger Ring, 61
"Animal Magnetism" Motor, 13

Black Hand, 71
Black Magic Art Show, 140
Blind Spot, 129
Boundless Menagerie, 132

Capillary Attraction, 102
Capturing Smoke in a Glass of Water, 58
Card Trick, 1,089, 41
Changing Two Liquids Into a Solid, 66
Chemical Novelties, 64
Classic Show (Black Magic), 140
Color Disc, 127
Color Changing Power, 63
Complete Magical Magnetism Act, 19

Dancing Doll, 7
Dancing Paper Dolls, 112
Darting Camphor, 65
Decapitated Princess Illusion, 137
Dry Water, 67

Eggs that Read, 59

Family-History Arithmetic, 34
Five Odd Figures, 37
Flash of Fire, 68
Funny Arithmetic, 35

Ghostly Faces, 69
Ghost Rings, 88
Guessing the Erased Number, 36

Hindu Climbing Balls, 108
How to Be a Lightning Calculator, 42
How to See Through a Solid Brick, 136
How to Tune a Guitar Without Using Your Ears, 96
How to Weigh "Nothing", 82
Human Calculator, 41
Human Magnet, 16

I Have Your Number, 44
Indestructible Card, 64
Invisible Coin Trick, 125
Is It Wine or Is It Water? 56

Jars of Mystery Vapor, 86

Levitated Needle, 4

Magical Humming Glass, 98
Magic Arrow, 11
Magic Flower Seeds, 62
Magic Mirror, 133
Magic Pipe, 116
Magic Sand of Arabia, 72
Magic Soda Fountain Act, 73
Magic Wound, 68
Magnetic Impulse, 24
Magnetic Motor, 5
Magnetic Table Lifting, 27
Magnetized Cane, 17
Magnetized Playing Cards, 25
Magnetized Poker, 2
Magnetized Stick, 23
Magnetized Strips of Paper, 21
Magnet Brush, 1
Making a Dry-Ice Fountain, 81
Making Water Boil from Freezing Cold, 83
Matchbox Monte Trick, 100
Miniaturized "Black Art", 143
More Novel Tricks with Dry Ice, 85
Musical Bottle, 94
Musical Glass, 94
Mysterious Number Nine, 50

Number Game, 49

Obedient Egg, 120

Periscope, 135
Prancing Horse, 117
Precarious Balancing, 118

Prism Rainbow, 126
Psychic Book Test, 38
Putting Candles Out by Magic, 82

"Quickie" Tricks with Dry Ice, 83

Racing Rain Drops, 104
Rainbow Water, 75
Red Hand, 70
Restless Mothball, 65
Revolving Pencil, 9
Revolving Snake, 113
Rubber-Band Snakes, 87

"Seebackascope", 134
Soft Drinks for Everyone, 88
Some Number Funnies, 46
Sound Magnifier, 97
Suspension, 121
Swinging Magnets, 3

"Thaumatrope", 128
Tricky Numbers, 30
Tricky Soap, 66

Water Mirror, 131
Whirling a Glass of Water Upside Down Without Spilling a Drop, 107
Which Hand Holds Which Coin? 38
Whirling Coat Hanger, 106
Wine and Water Trick, 53

Visible Sound Vibrations, 93

INDEX

Acid reactions, 56
Acoustics, magic with, 92–93
　Aeolian harp, 97–98
　box, music, 94
　glass, humming, 98–99
　glass, musical, 94–96
　guitar, tuning without
　　using ears, 96
　matchbook monte trick,
　　100–102
　sound magnifier, 97
　sound vibrations, visible,
　　93–94
Aeolian harp, 97–98
Age and loose change trick,
　32–33
Alchemy, 51
Alkaline reactions, 56
Alum, creating chemical tube,
　61
Amplification. *See* Acoustics
Animal magnetism, 13–15
Antigravity, 115
Arithmetic, funny, 35–36
Arrow, magic, 11–13, 14
Attraction, 115
　capillary, 102–105
　magnetism, 3–4
Audio misdirection, 99, 101

Balancing, precarious, 118–
　120

Balls, Hindu climbing, 108–
　111
Black Art Magic Show, 140
　classic show, 140–143
　miniaturized chamber,
　　143–150
Blind spot, 129–130
Bottle
　lifting with a straw, 122–
　　123
　musical, 94
Brick, how to see through a
　solid, 136–137
Bridge, self-supporting with
　knives, 123
Brittleness, 114
Brush, magnet, 1–2
Burning of fuel, 79

Calculator
　human, 41–42
　lightning, 42–44
Candles, putting out by
　magic, 82–83
Camphor, darting, 65
Cane, magnetized, 17–18
Capillary attraction, 102–105
Carbon dioxide, frozen. *See*
　Dry Ice
Card
　indestructible, 64

playing, magnetized, 25–27

trick, 1, 089, 41

Center of gravity, 115–121

Centrifugal force, 106

balls, Hindu climbing, 108–111

coat hanger, whirling, 106–107

glass, whirling, 107–108

Change, loose, and age trick, 32–33

Chemicals, handling, 51, 52

Chemicals, magic with, 74–75

camphor, darting, 65–66

card, indestructible, 64

faces, ghostly, 69

fire, flash of, 68

hand, red and black, 70–71

liquids into a solid, 66

mothball, restless, 65

sand, magic, 72–73

soap, tricky, 66–67

water, dry, 67

wound, magic, 68–69

Chemistry, magic with, 51–52

chemical magic entertainment, 74–75

chemical novelties, 64–73

color changing power, 63–64

eggs that read, 59–61

finger ring, 61

rainbow waters, 75–77

smoke in glass of water, 58

soda fountain set, magic, 73–74

wine and water trick, 53–56

wine or water, 56–57

Children, teaching the art of magic to, 151–155

Clock reaction, 57, 58

Coat hanger, whirling, 106–107

Coins

trick, invisible, 125

which hand holds which, 38

Color

black, invisibility of, 140–150

changing power, 63–64

disc, 127–128

spectrum, 126

Contraction, 102

Cube root extraction, 42, 43–44

Decapitated Princess illusion, 137–140

Decay, principle of, 79

Density of solutions, principle of, 60

Divisibility, 114

Doll

dancing, 7–9, 10

dancing paper, 112–113

Dry ice, magic with, 78–81

candles, putting out by magic, 82–83

fountain, making a, 81

rings, ghost, 88–91

snakes, rubber-band, 87

soft drinks, 88
tricks, 83–86
vapor, jars of mysterious,
    86
water, making boiling, 83
weighing "nothing," 82
Ductility, 114
Duration of visual
    impressions, 128–129

Eggs
    obedient, 120–121
    reading, 59–60
Elasticity, 114
Electricity, static, 11–12, 21
Erased number, guessing the,
    36–37
Evaporation, 79
Eye, blind spot, 129–130

Faces, ghostly, 69
Family history arithmetic,
    34–35
Fermentation, 79
Fire, flash of, 68
Fountain
    dry-ice, 81
    soda, magic, 73–74
Flotation, 102
Flower seeds, magic, 62–63
Force, 40, 114–115
    attraction, magnetic, 3–4
    centrifugal, 106–111
    repulsion, magnetic, 3–4
Fuel, burning of, 79

Gas reactions, 65
Ghostly faces, 69

Ghost rings, 88–91
Glass
    magical humming, 98–99
    musical, 94–96
    smoke in, 58
    whirling water without
        spilling, 107–108
Gravity, 115
    balancing, precarious, 118–
        120
    egg, obedient, 120–121
    horse, prancing, 117–118
    pipe, magic, 116–117
Guitar, tuning without using
    your ears, 96

Hand
    black, 71
    red, 70–71
Harp, Aeolian, 97–98
Hearing. *See* Acoustics
Heat, 111
    magnetism and, 5–6
    paper dolls, dancing, 112–
        113
    snake, revolving, 113–114
Hindu climbing balls, 108–
    111
Horse, prancing, 117–118
Human calculator, 41–42
Human magnet, 16–17
Humming glass, magical, 98–
    99

Illusions, 152
Impulse, magnetic, 24–25

Knives, self-supporting bridge
    with, 123

Levitation, 4, 115
Light. *See* Optics
Liquids
    capillary attraction, 102–
        105
    changing into a solid, 66

Magical act, complete
    magnetism, 19–21
    impulse, magnetic, 24–25
    newspaper, magnetized
        strips of, 21–22
    playing cards, magnetized,
        25–27
    stick, magnetized, 23–24
    table lifting, magnetic, 27–
        29
Magic show
    black art, 140–150
    performance of, 20, 153–
        155
Magic, teaching children the
    art of, 151–155
Magic with
    chemistry, 51–77
    dry ice, 78–91
    magnetism, 1–29
    numbers, 30–50
    physics, 92–150
Magnetism, magic with
    arrow, magic, 11–13
    cane, magnetized, 17–18
    doll, dancing, 7–9
    human magnet, 16–17
    magical act, complete, 19–
        29
    motor, "animal
        magnetism," 13–15

    motor, magnetic, 5–6
    pencil, 9–10
Malleability, 114
Matchbook monte trick, 100–
    102
Matches
    hungry, 102–103
    lifting, 121–122
Matter, properties of, 114
Mechanics, 114–115
    gravity, 115–121
    suspension, 121–124
Menagerie, boundless, 132
"Mercury hammer," 85
Mirror
    brick, seeing through solid,
        136—137
    decapitated princess
        illusion, 137–140
    magic, 133–134
    menagerie, boundless, 132
    periscope, 135
    seebackascope, 134–135
    water, 131
Mobility, 114
Mothball, restless, 65
Motion
    laws of, 114
    obliquity of, 114
    points of least, 96
Motor
    animal magnetism, 13–15
    magnetic, 5–6

Needle, levitated, 4
Newspaper, magnetized strips
    of, 21–22

Nine, mysteries of number, 50
Nodes, 96
Numbers, magic with
add a number, 32
age and loose change trick, 32–33
card trick, 41
coin, which hand holds which, 38
erased number, guessing the, 36–37
family-history arithmetic, 34–35
five odd figures, 37–38
funnies, 46–48
funny arithmetic, 35–36
game, 49
guessing, 31, 44–45
human calculator, 41–44
number nine, 50
psychic book test, 38–40
tricky numbers, 30–31

Obliquity of motion, 114
Odd figures, five, 37–38
Optics, 124–125
blind spot, 129–131
brick, how to see through, 136–137
coin trick, invisible, 125
color disc, 127–128
decapitated princess illusion, 137–140
menagerie, boundless, 132
mirror, magic, 133–134
mirror, water, 131
periscope, 135

prism, rainbow, 126
seebackascope, 134–135
thaumatrope, 128–129
Oxidation, 78–79

Paper dolls, dancing, 112–113
Pencil, revolving, 9–10
Periscope, 135
Physics, magic with, 92
acoustics, 92–102
black art magic show, 140–150
capillary attraction, 102–105
centrifugal force, 106–111
heat, 111–114
mechanics, 114–124
optics, 124–140
Pipe, magic, 116–117
Playing cards, magnetized, 25–27
Points of least motion, 96
Poker, magnetized, 2
Poles, magnetic, 1–2
Primary colors, 126
Prism, 126–128
Psychic book test, 38–40

Rainbow waters, 75–77
Reflection, 131, 132, 133–134
Refraction, 125
Repulsive force, magnet, 3–4
Resolution of forces, 115
Respiration, 79
Rings
amazing finger, 61

ghost, 88–91
Routining, 20
Rubber bands
  freezing, 85
  snakes, 87

Sand, magic, 72–73
Saturation, 66
Secondary colors, 126
"Seebackascope," 134–135
Seeds, magic flower, 62–63
Smoke, capturing in a glass of
  water, 58
Snake, revolving, 113–114
Soap, tricky, 66–67
Solidity, 114
Soft drinks, making, 73–74,
  88
Soda fountain act, magic, 73–
  74
Solid from liquids, 66
Solutions, chemical, 52
Sound, 92–93
  magnifier, 97
  misdirection, 99, 101
  waves, 93–94
  see also Acoustics
Spectrum, color, 126
Static electricity, 11–12, 21
Stick, magnetized, 23–24
Straw, lifting a bottle with a,
  122–123
Superficial tension of the
  liquid, 102
Supernatural, interest in the,
  152

Suspension, 121–124
Swinging magnets, 3–4

Table lifting, magnetic, 27–
  29
Teaching children the art of
  magic, 151–155
Tenacity, 114
"Thaumatrope," 128–129
Tuning a guitar without using
  your ears, 96
Tuning fork, 94

Uphill, rolling an object,
  123–124

Vapor, jars of mystery, 86
Ventral segments of vibration,
  96
Vibration
  ventral segments of, 96
  visible sound, 93–94
Vision. *See* Optics

Water
  boiling from freezing cold,
    83
  drops, racing, 104–105
  dry, 67
  glass of, whirling without
    spilling, 107–108
  rainbow, 75–77
Weighing nothing, 82
Wine and water trick, 53–56
  is it wine or water?, 56–57
Wound, magic, 68–69